Springer Theses

Recognizing Outstanding Ph.D. Research

Aims and Scope

The series "Springer Theses" brings together a selection of the very best Ph.D. theses from around the world and across the physical sciences. Nominated and endorsed by two recognized specialists, each published volume has been selected for its scientific excellence and the high impact of its contents for the pertinent field of research. For greater accessibility to non-specialists, the published versions include an extended introduction, as well as a foreword by the student's supervisor explaining the special relevance of the work for the field. As a whole, the series will provide a valuable resource both for newcomers to the research fields described, and for other scientists seeking detailed background information on special questions. Finally, it provides an accredited documentation of the valuable contributions made by today's younger generation of scientists.

Theses are accepted into the series by invited nomination only and must fulfill all of the following criteria

- They must be written in good English.
- The topic should fall within the confines of Chemistry, Physics, Earth Sciences, Engineering and related interdisciplinary fields such as Materials, Nanoscience, Chemical Engineering, Complex Systems and Biophysics.
- The work reported in the thesis must represent a significant scientific advance.
- If the thesis includes previously published material, permission to reproduce this must be gained from the respective copyright holder.
- They must have been examined and passed during the 12 months prior to nomination.
- Each thesis should include a foreword by the supervisor outlining the significance of its content.
- The theses should have a clearly defined structure including an introduction accessible to scientists not expert in that particular field.

More information about this series at http://www.springer.com/series/8790

Raphaëlle D. Haywood

Radial-velocity Searches for Planets Around Active Stars

Doctoral Thesis accepted by
the University of St Andrews, UK

Author
Dr. Raphaëlle D. Haywood
Harvard College Observatory
Cambridge, MA
USA

Supervisor
Prof. Andrew Collier Cameron
School of Physics and Astronomy
University of St Andrews
St Andrews
UK

ISSN 2190-5053 ISSN 2190-5061 (electronic)
Springer Theses
ISBN 978-3-319-82310-2 ISBN 978-3-319-41273-3 (eBook)
DOI 10.1007/978-3-319-41273-3

© Springer International Publishing Switzerland 2016
Softcover reprint of the hardcover 1st edition 2016
This work is subject to copyright. All rights are reserved by the Publisher, whether the whole or part of the material is concerned, specifically the rights of translation, reprinting, reuse of illustrations, recitation, broadcasting, reproduction on microfilms or in any other physical way, and transmission or information storage and retrieval, electronic adaptation, computer software, or by similar or dissimilar methodology now known or hereafter developed.
The use of general descriptive names, registered names, trademarks, service marks, etc. in this publication does not imply, even in the absence of a specific statement, that such names are exempt from the relevant protective laws and regulations and therefore free for general use.
The publisher, the authors and the editors are safe to assume that the advice and information in this book are believed to be true and accurate at the date of publication. Neither the publisher nor the authors or the editors give a warranty, express or implied, with respect to the material contained herein or for any errors or omissions that may have been made.

Printed on acid-free paper

This Springer imprint is published by Springer Nature
The registered company is Springer International Publishing AG Switzerland

*A mon père,
modèle de force, courage,
patience et détermination*

Supervisor's Foreword

Raphaëlle Haywood's thesis develops powerful new methods for detecting reflex orbital motions of solar-type stars hosting extra-solar planets, in the presence of stellar magnetic activity.

During the first two decades of exoplanet research, the sensitivity of radial-velocity spectrometers improved exponentially, at the rate of one order of magnitude per decade. After 2011, however, the detection threshold stalled at orbital velocity amplitudes close to 1 m s^{-2}. Although the new generation of high-accuracy radial-velocity spectrometers is capable of detecting velocity shifts an order of magnitude smaller than this, the photospheric physics of the host stars themselves is now the limiting factor.

High-resolution images of our nearest star, the Sun, show that its surface churns with convective flows on a wide range of length scales. The solar granulation in particular is suppressed in the regions of high magnetic field strength that surround sunspot groups and make up the wider solar magnetic network. As the Sun rotates, sunspots and active regions pass in and out of view, modulating the Sun's apparent radial velocity by several m s^{-2}. Space-borne observations of other Sun-like stars show similar patterns of modulation in brightness, which are closely related to the velocity variations that plague efforts to determine the masses of their small planets.

The need to overcome this barrier has become more acute since the advent of space-based photometry missions such as CoRoT and Kepler, which have detected the transits of planets down to the size of the Earth and even smaller. To have any hope of distinguishing rocky Earth analogues from mini-Neptunes with low-density ice mantles, their masses must be found by measuring the orbital reflex motions of their host stars and disentangling them from the higher amplitude stellar activity signal.

Using the High-Accuracy Radial-velocity Planet Searcher (HARPS) on the European Southern Observatory's 3.6-m telescope at La Silla, Raphaëlle Haywood conducted observations of the compact planetary system orbiting the magnetically active star CoRoT-7, simultaneously with photometry from the CoRoT spacecraft. Her analysis of these data sets demonstrates that the radial-velocity variations arise mainly from suppression of photospheric convection by magnetic fields. A key result of Haywood's work on CoRoT-7 was the recognition that while stellar active regions come and go, a true planetary signal remains constant in phase and amplitude. Her work provides the first practical demonstration that Gaussian process regression is adept at teasing them apart, given a sufficiently long and well-sampled data train.

Haywood's second major achievement was to carry out the first systematic campaign of radial-velocity observations of the Sun using the HARPS instrument, using integrated sunlight scattered from the surface of a bright asteroid. She used data from the Solar Dynamics Observatory to identify the types of solar surface activity that drive the full-disc velocity variations. She demonstrated that the suppression of convective blueshift in solar active regions, and the velocity modulation caused by dark spots and bright faculae rotating across the face of the Sun, was directly measurable from the SDO images. She found them to be an excellent predictor of the Sun-as-a-star radial-velocity fluctuations measured over two solar rotations with the HARPS instrument.

The clarity of Haywood's writing makes her thesis popular with researchers in the field seeking to master and adopt the state-of-the-art statistical methods that she employed. These include Gaussian-process regression for modelling the correlated signals arising from evolving active regions on a rotating star and Bayesian model selection methods for distinguishing genuine planetary reflex motion from false positives arising from stellar magnetic activity.

Her study represents a significant step towards measuring the masses of potentially habitable planets orbiting Sun-like stars with solar-like magnetic activity. The techniques she developed are influential in the design of new observing strategies that allow intrinsic stellar variability to be fully characterised and separated from planetary motion, using the data analysis methods described in the thesis. Although the first mass measurement of a true Earth analogue orbiting in the habitable zone of a Sun-like star is still some way off, the methods pioneered in this thesis represent an influential milestone along the journey.

St Andrews, UK
February 2016

Prof. Andrew Collier Cameron

Preface

Since the discovery of the first planet orbiting another star than our Sun, just over twenty years ago, hundreds of new extra-solar planets have been identified, and thousands of more discoveries are awaiting confirmation. The first exoplanets that were detected had sizes similar to those of Jupiter and Saturn, the giants in our solar system. In recent years, instrument precision and telescope power have improved so much that discovering and characterising planets as small as the Earth is now a reality. The search for worlds similar to our own is one of the fastest growing fields in astronomy; it is a young and exciting field and captivates the interest of the public like no other.

One of the most successful ways to find extra-solar planets is to look for stars that wobble. As a planet orbits around its parent star, it exerts a tiny pull on the star. This causes the starlight to periodically stretch and compress, making the star appear redder and bluer. This effect, known as the Doppler shift, is the same effect that makes the siren of an ambulance sound high-pitched then low-pitched as it drives past. These minuscule changes in the colour of the star's light, which reflect the variations of the star's velocity along our line of sight, can be detected by current state-of-the-art spectrographs.

There are still several challenges to be overcome in the quest for other Earths. One major difficulty arises from the intrinsic magnetic activity of the host stars themselves. Indeed, the correlated noise that arises from their natural radial-velocity variability can easily mimic or conceal the orbital signals of super-Earth and Earth-mass exoplanets, and there is currently no reliable method to untangle the signal of a planet from this stellar "noise".

The work I undertook as part of my thesis was intended to tackle this issue via a twofold approach. First, I developed an intuitive and robust data analysis framework in which the activity-induced variations are modelled with a Gaussian process that has the frequency structure of the photometric variations of the star, thus allowing me to determine precise and reliable planetary masses (Chap. 2); and second, I explored the physical origin of stellar-induced Doppler variations through the study of our best-known star, the Sun (Chap. 4).

I applied my new data-modelling technique to three recently discovered planetary systems: CoRoT-7, Kepler-78, and Kepler-10 (Chap. 3). I determined the masses of the transiting super-Earth CoRoT-7b and the small Neptune CoRoT-7c to be $4.73 \pm 0.95\,M_\oplus$ and $13.56 \pm 1.08\,M_\oplus$, respectively. The density of CoRoT-7b is 6.61 ± 1.72 g cm^{-3}, which is compatible with a rocky composition. I carried out Bayesian model comparison to assess the nature of a previously identified signal at 9 days and found that it is best interpreted as stellar activity. Despite the high levels of activity of its host star, I determined the mass of the Earth-sized planet Kepler-78b to be $1.76 \pm 0.18\,M_\oplus$. With a density of $6.2^{+1.8}_{-1.4}$ g cm^{-3}, it is also a rocky planet. I found the masses of Kepler-10b and Kepler-10c to be $3.31 \pm 0.32\,M_\oplus$ and $16.25 \pm 3.66\,M_\oplus$, respectively. Their densities, of $6.4^{+1.1}_{-0.7}$ g cm^{-3} and 8.1 ± 1.8 g cm^{-3}, imply that they are both of rocky composition—even the 2 Earth-radius planet Kepler-10c!

In parallel, I deepened our understanding of the physical origin of stellar radial-velocity variability through the study of the Sun, which is the only star whose surface can be imaged at high resolution. I found that the full-disc magnetic flux is an excellent proxy for activity-induced radial-velocity variations; this result may become key to breaking the activity barrier in coming years.

I also found that in the case of CoRoT-7, the suppression of convective blueshift leads to radial-velocity variations with an RMS of 1.82 m s^{-1}, while the modulation induced by the presence of dark spots on the rotating stellar disc has an RMS of 0.46 m s^{-1}. For the Sun, I found these contributions to be 2.22 m s^{-1} and 0.14 m s^{-1}, respectively. These results suggest that for slowly rotating stars, the suppression of convective blueshift is the dominant contributor to the activity-modulated radial-velocity signal, rather than the rotational Doppler shift of the flux blocked by starspots.

Gaining a deeper understanding of the physics at the heart of activity-driven RV variability will ultimately enable us to better model and remove this contribution from RV observations, thus revealing the planetary signals.

Cambridge, MA, USA   Dr. Raphaëlle D. Haywood

Acknowledgements

First, I would like to thank my supervisor Andrew for his unwavering support throughout my Ph.D. I enjoyed our numerous debug sessions, which you always made time for despite your Head of School duties. Your boundless enthusiasm has made research more fun than I would have ever imagined! Thank you also for sending me on all those conferences and observing trips—I have discovered much more than exoplanets in the past three years!

I am also very grateful to my first ever journal referee, Suzanne Aigrain, who introduced me to the power of Gaussian processes—a cornerstone in my thesis work.

I am indebted to everyone in my office, particularly Will, Laura, Milena, and especially Claire, for listening to my incessant rants, both happy and angry, and for cheering me up when I needed it. For this, I should also thank the postdocs in the office next door and the astronomy group as a whole. I hope the Monday cake tradition and the long coffee breaks continue!

I am grateful to Joe and Will for all their Fortran-related help, to Annelies for answering my random science questions and for post-telecon rants, and to Guillaume for providing me extremely useful advice for my postdoc applications.

Joe, thank you for the many evenings spent at your house back in my first year and for the fun we had at conferences together—there will be more soon! Jack, thank you for the endless hours we spent talking about everything and nothing, sometimes stretching far into the night, or sitting on a tree over a river. Rim, I miss our late lunches at Pizza Express, your crazy talk (your serious talk too) and all our laughs about silly things! Now that I am done with this thesis I really ought to visit you in Sicily.

João, you kept me sane during the final months of writing-up. You stood by me even in the times when I was as intolerable as only my brother James knows—I am sure he will tell you that is quite an achievement. Also, you proofread pretty much the entirety of this thesis. Thank you!

Finally, I wish to give an enormous thank you to my family, who has always encouraged and supported me. None of this would have been possible without the very frequent phone conversations with my Dad, and his constant stream of advice and encouragement. I am grateful for the regular Sunday phone calls to my grandma, and the comforting (and sometimes hilarious) chats with my sisters Adèle and Lucie and my brothers James and Germain. I am also very much indebted to Anne-Sophie for her support. I look forward to another great family holiday together!

Contents

1 **Introduction: The Hunt for Extra-Solar Planets** 1
  References 9

2 **Stellar Activity as a Source of Radial-Velocity Variability** 13
  2.1 Magnetic Activity and Its Manifestations 13
      2.1.1 Minutes: Oscillations 15
      2.1.2 Minutes: Flares and Coronal Mass Ejections 16
      2.1.3 Minutes to Hours: Granulation 16
      2.1.4 Days and Longer: Gravitational Redshift 17
      2.1.5 Stellar Rotation Period: Spots, Faculae
            and Plage Regions............................... 18
      2.1.6 Decades: Magnetic Cycles 23
      2.1.7 Timescales: Summary 25
  2.2 Existing Treatments for Activity-Induced RV Variations...... 25
      2.2.1 Spectroscopic Activity Indicators.................. 26
      2.2.2 Nightly Offsets Method 28
      2.2.3 Harmonic Decomposition 30
      2.2.4 Pre-whitening 31
      2.2.5 The FF' Method 31
      2.2.6 Existing Methods: Summary........................ 32
  2.3 RV Target Selection Based on Photometric Variability......... 33
      2.3.1 Preliminary Target Selection Criteria 33
      2.3.2 Generalised Lomb–Scargle Periodograms and
            Autocorrelation Functions......................... 34
      2.3.3 Selection Criteria for "Magnetically Manageable"
            Stars 40
  2.4 Concluding Note: From Photometric to Radial-Velocity
      Variations... 41
  References 41

xiii

3 A Toolkit to Detect Planets Around Active Stars ... 45
3.1 Gaussian Processes ... 45
- 3.1.1 Definition ... 46
- 3.1.2 Covariance Matrix K ... 50
- 3.1.3 Covariance Function $k(t,t')$ ... 53
- 3.1.4 Temporal Structure and Covariance ... 54
- 3.1.5 Gaussian Processes for Stellar Activity Signals ... 55
- 3.1.6 Determining the Hyperparameters θ_j ... 56
- 3.1.7 Constructing the Covariance Matrix K ... 56
- 3.1.8 Fitting Existing Data and Making Predictions ... 57
- 3.1.9 A Word of Caution ... 58
- 3.1.10 Useful References ... 58

3.2 Monte Carlo Markov Chain (MCMC) ... 59
- 3.2.1 Modelling Planets ... 59
- 3.2.2 Modelling Stellar Activity ... 60
- 3.2.3 Total RV Model ... 62
- 3.2.4 Choice of Priors ... 62
- 3.2.5 Fitting Procedure ... 64
- 3.2.6 Care Instructions ... 65

3.3 Model Selection with Bayesian Inference ... 66
- 3.3.1 Bayes' Factor ... 67
- 3.3.2 Posterior Ordinate ... 67
- 3.3.3 Marginal Likelihood ... 68

References ... 68

4 Application to Observations of Planet-Hosting Stars ... 71
4.1 CoRoT-7 ... 72
- 4.1.1 History of the System ... 72
- 4.1.2 Observations ... 75
- 4.1.3 Preliminary Periodogram Analysis ... 77
- 4.1.4 MCMC Analysis ... 78
- 4.1.5 Results and Discussion ... 80
- 4.1.6 Summary ... 88

4.2 Kepler-78 ... 89
- 4.2.1 History of the System ... 89
- 4.2.2 Observations ... 92
- 4.2.3 MCMC Analysis ... 93
- 4.2.4 Results and Discussion ... 95
- 4.2.5 Summary ... 98

4.3 Kepler-10 ... 98
- 4.3.1 History of the System ... 98
- 4.3.2 Observations ... 100
- 4.3.3 MCMC Analysis ... 102
- 4.3.4 Results and Discussion ... 103
- 4.3.5 Summary ... 107

	4.4	Summary and Future Plans	108
		4.4.1 Determining the Bulk Densities of Transiting Exoplanets	108
		4.4.2 Assessing the Reliability of the Gaussian Process Framework for Exoplanet Mass Determinations	108
		4.4.3 Concluding Note	110
	References		111
5	**An Exploration into the Radial-Velocity Variability of the Sun**		**113**
	5.1	Previous Studies on the Intrinsic RV Variability of the Sun	114
	5.2	HARPS Observations of Sunlight Scattered Off Vesta	115
		5.2.1 HARPS Spectra	115
		5.2.2 Solar Rest Frame	116
		5.2.3 Relativistic Doppler Effects	116
		5.2.4 Sources of Intra-Night RV Variations	118
		5.2.5 Time Lag Between Vesta and SDO Observations	120
	5.3	Pixel Statistics from SDO/HMI Images	120
		5.3.1 Spacecraft Motion	121
		5.3.2 Solar Rotation	121
		5.3.3 Flattened Continuum Intensity	122
		5.3.4 Unsigned Longitudinal Magnetic Field Strength	122
		5.3.5 Surface Markers of Magnetic Activity	123
	5.4	Reproducing the RV Variations of the Sun	128
		5.4.1 Total RV Model	128
		5.4.2 Relative Importance of Suppression of Convective Blueshift and Sunspot Flux Deficit	129
		5.4.3 Zero Point of HARPS	130
	5.5	Towards Better Proxies for RV Observations	130
		5.5.1 Disc-Averaged Observed Magnetic Flux $|\hat{B}_{obs}|$	130
		5.5.2 Correlations Between RV and Activity Indicators	131
	5.6	Summary	132
	References		133
6	**Conclusion: Next Steps and Aims for the Future**		**135**
	References		138
Index			**139**

Chapter 1
Introduction: The Hunt for Extra-Solar Planets

> *In future, children won't perceive the stars as mere twinkling points of light: they'll learn that each is a 'Sun', orbited by planets fully as interesting as those in our Solar system.*
>
> Sir Martin Rees, 2003

Since the dawn of civilisation we have looked up to the stars, wondering whether other worlds exist and what they might look like. In the last few decades, developments in instrumentation and observation techniques have led to revolutionary discoveries: we now know that planets revolving around other stars than our Sun exist, and better still, they appear to be very common.

Early searches In the 1940s, a few independent exoplanet discovery claims were made (Strand 1943; Reuyl and Holmberg 1943), based on perturbations in the astrometric motions of their host stars. There was also a flurry of interest in astrometric detection in the late 1960s, when van de Kamp (1969) claimed the detection of two planets orbiting Barnard's star. Although these findings were soon proved wrong, they sparked new appeal in the astronomy community at the time. The idea that we might be able to detect the radial motion of a star induced by the gravitational tug of orbiting planets was proposed by Struve (1952), and several radial-velocity (RV) monitoring surveys were initiated, including those of Campbell and Walker (1985), Latham et al. (1989) and Marcy and Butler (1994). This technique has proved very successful at detecting and confirming exoplanets since then, and has yielded some of the most exciting results in this new field of astronomy.

Wolszczan and Frail (1992) reported on the first detection of two Earth-mass planets orbiting a pulsar, based on variations in the timing of its pulses. In 1995, Mayor and Queloz (1995) announced the discovery of 51 Peg b, the first planet-mass companion of a Sun-like star. It was found through RV observations taken with the ELODIE spectrograph (Baranne et al. 1996), mounted on the 1.93-m telescope at the Observatoire de Haute-Provence (France). 51 Peg b has half the mass of Jupiter, but orbits its host star once every 4 days: it is much closer to its star than Mercury, which orbits the Sun every 88 days.

Fig. 1.1 Yesterday's discoveries and today's challenges: RV variations of WASP-8, host to a hot-Jupiter in an 8-day orbit recorded with CORALIE and HARPS (Queloz et al. 2010) (*blue*), and HARPS RV variations of CoRoT-7, an active star host to a super-Earth in an 0.85-day orbit and a small Neptune at 3.65 days (Haywood et al. 2014) (*red*). The *horizontal* and *vertical* scales are the same for both RV curves

An army of planet-hunting spectrographs was formed including SOPHIE, to replace ELODIE on the 1.93 m (Perruchot et al. 2008) and HARPS (High Accuracy Radial-Velocity Planet Searcher, Mayor et al. 2003), commissioned in 2003 on the 3.6 m telescope at La Silla, Chile. The first planet-hunting spectrographs, including ELODIE, HIRES, mounted on Keck I at Mauna Kea observatory, Hawaii (Vogt et al. 1994), and CORALIE, mounted on the Euler telescope at La Silla, Chile (Queloz et al. 2000) had an RV precision of about $15\,\text{m}\cdot\text{s}^{-1}$ (see Perryman 2011, p. 24. for a complete list of planet-hunting spectrographs). Over the years, improved calibration techniques (see Mayor et al. 2014; Pepe et al. 2014a and references therein) have pushed the RV sensitivity down by an order of magnitude. HARPS is able to detect RV signals with amplitudes as low as $1\,\text{m}\cdot\text{s}^{-1}$ (Queloz et al. 2001b), and has paved the way towards the discovery of Neptune- and super-Earth-mass planets through new blind RV planet surveys (see Fig. 1.1).

From RV monitoring to planet detection The radial velocity of a star is defined as "the component of its motion along the line of sight of the observer" (Murdin 2002). The presence of a planet exerts a gravitational pull on the star that causes it to wobble by tiny amounts around their common centre of mass, as shown in Fig. 1.2a. The light we receive from the star is slightly blueshifted or redshifted as the star gets pulled towards or away from us (Fig. 1.2b).

As it reaches the observer, light with a wavelength λ_{obs} has undergone a relativistic Doppler shift relative to when it was initially emitted by the star as λ_{rest} (the wavelength of a spectral line in a rest frame). Each line is thus shifted by an amount:

$$\Delta\lambda = \lambda_{\text{obs}} - \lambda_{\text{rest}}. \tag{1.1}$$

It is possible to show that the radial component of the velocity of the star is proportional to this wavelength shift:

$$v_{\text{radial}} \approx \frac{\Delta\lambda}{\lambda_{\text{rest}}} c, \tag{1.2}$$

1 Introduction: The Hunt for Extra-Solar Planets

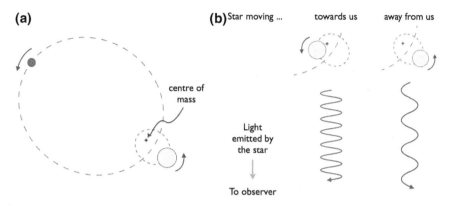

Fig. 1.2 The radial-velocity method: as a planet orbits its host star, the star wobbles around their common centre of mass (*panel* **a**). As a result, the starlight is Doppler shifted (*panel* **b**)

where c is the speed of light (refer to Perryman (2011, p. 16), and references therein for the full derivation of this equation).

Monitoring the RV of a star in time allows us to detect any variations induced by the orbit of a planet. A few examples of existing and hypothetical planets and their expected RV semi-amplitudes are listed in Table 1.1.

The mass m of the planet can be derived from the semi-amplitude K and period P of the signal, if we know the stellar mass $M_\star$. The mass function $f(m)$ is given by:

$$f(m) = \frac{m^3 \sin i}{(M_\star + m)^2} = \frac{K^3 P}{2\pi G} (1 - e^2)^{3/2}, \tag{1.3}$$

where i is the planet's orbital inclination and G is the gravitational constant (the full derivation of this equation can be found in Perryman 2011 or Hilditch 2001).

Ground-based photometric surveys The surprising discovery of 51 Peg b was soon followed by others, through photometric surveys such as the Hungarian Automated Telescope Network, operational since 2001 (HATNet, Bakos et al. 2004),

Table 1.1 Approximate RV semi-amplitudes expected from some of the planets of our solar system at a range of distances (values obtained from the Wikipedia page "Doppler spectroscopy", in March 2015)

Planet mass	Distance to star (AU)	Orbital period	K (m·s^{-1})
Jupiter (317 $M_\oplus$)	5	5 years	12
Jupiter	1	1 year	28
Neptune (17 $M_\oplus$)	0.1	36 days	5
Super-Earth (5 $M_\oplus$)	0.1	36 days	1.4
Earth	1	1 year	0.09

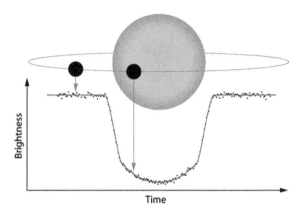

Fig. 1.3 Transit of WASP-10b: the planet casts its shadow upon the stellar disc when it passes in front of it, thereby reducing the total brightness observed. Image credit: John Johnson

and the Wide-Angle Search for Planets initiated in 2004 (WASP, Pollacco et al. 2006; Cameron et al. 2009). These arrays of small robotic telescopes monitor the brightness variations of hundreds of thousands of stars at a time, looking for tiny, periodic dips of less than 1 % in the star's light, which may be caused by a planet crossing the disc of its host star. Together, WASP and HATNet have found over 200 transiting planets as of March 2015.

One of the most common ways to detect extra-solar planets is to look for the periodic dimming of the host star as a planet passes in front of it relative to the observer, as illustrated in Fig. 1.3. Transit events are very unlikely as they require the observer, planet and host star to be very well aligned. If the radius of the star and the orbital eccentricity are known, then the radius of the planet can be inferred (see Seager and Mallen Ornelas 2003). If our planetary system contains one or more transiting planets we can assume i approximately equal to 90°. This is usually a reasonable assumption for all the planets in a compact system since over 85 % of observed compact planetary systems containing transiting super-Earths and Neptunes are thought to be coplanar within 3° (Lissauer et al. 2011b).

Combining transit and RV observations together yields a complete set of planetary and orbital parameters. The bulk density of the planet can then be inferred from its mass and radius, allowing us to take a first guess at its structure and composition.

The space revolution After the turn of the century, several photometric satellites were launched into space to look for transits of small super-Earth- and Earth-size planets without the disruptive twinkling induced by the Earth's atmosphere. The European CoRoT[1] satellite (Baglin and Team 1998; Auvergne et al. 2009), launched at the end of 2006 was the first space mission dedicated, in part, to the detection of

---

[1]COnvection, ROtation et Transits planétaires.

1 Introduction: The Hunt for Extra-Solar Planets

exoplanets. It discovered 32 validated planets to date (see a recent review by Hatzes 2014), most of them giant gas planets, but also several small rocky planets, including CoRoT-7b, the first transiting rocky planet ever discovered (Léger et al. 2009).

In addition to finding transiting exoplanets, the CoRoT satellite, together with the Canadian space mission MOST[2] (Matthews et al. 2000) launched in 2003 and the *Kepler* mission (launched in 2009) sparked huge advances in the field of asteroseismology. Sun-like stars constantly pulsate due to the numerous acoustic oscillations bouncing within their interiors; by monitoring the amplitude and frequency structure of the flickering induced by these oscillations, we can probe stellar interiors and characterise stars with unprecedented accuracy and precision (see Campante 2015, Chaplin et al. 2014 and Kjeldsen et al. 2010 among others). In particular, asteroseismology provides very accurate and precise measurements of radii and masses of stars, which are essential to characterise extra-solar planets (see Campante 2015 and references therein).

In addition to these missions, the Gaia spacecraft was launched at the end of 2013 by ESA. It will measure fundamental parameters (including distance, radius and effective temperature) of about 1 billion stars, which amounts to 1 % of the stars in our Galaxy (Lindegren 2009). It is expected to detect 5000 transiting exoplanets via photometric monitoring as well as a further 2000 exoplanets via astrometric measurements, which should allow it to detect every Jupiter-mass planet with orbital periods between 1.5 and 9 years (de Bruijne 2012; Sozzetti 2011).

The *Kepler* space mission funded by NASA (Borucki et al. 2011; Koch et al. 2010) prompted an explosion in exoplanet discoveries: as of March 2015, over 1000 planets have been confirmed, and another 4200 candidates are awaiting further investigation. Because of their sheer number and that many of them are too faint for ground-based telescopes, *Kepler* candidates cannot all be confirmed via RV follow-up observations or transit timing variations induced by gravitational interactions between planets in multiple systems. Instead, a large number of candidates are now elevated to planet status via statistical validation (Rowe et al. 2014; Lissauer et al. 2014b; Torres et al. 2011). In this procedure, the likelihood of a planet nature is weighted against other possible phenomena such as a grazing eclipsing stellar binary, blend with a background binary system, instrumental effects, etc.

Among its most notable discoveries, *Kepler*'s first Earth-size rocky planet was Kepler-10b (Batalha et al. 2011); a rocky Neptune-mass companion Kepler-10c was confirmed soon after (Fressin et al. 2011). I determine the masses of both planets using HARPS-N RV data in Chap. 4. The first system characterised via transit timing variations was Kepler-9, a system of two giants (Holman et al. 2010); see also Kepler-36, a curious system because its two transiting planets have very different densities (Carter et al. 2012). Kepler-11 was found to host 6 small transiting planets, 5 of which have orbital periods between 10 and 47 days (Lissauer et al. 2011a, 2013); this was the first of many multiple compact systems. *Kepler* also found several circumbinary planets, including Kepler-16b (Doyle et al. 2011), Kepler-47b and c (Orosz et al. 2012b) and the Neptune-size Kepler-38b (Orosz et al. 2012a). The

---

[2]Microvariability and Oscillations of STars.

first planet discovered with the same radius and mass as the Earth was Kepler-78b (Sanchis-Ojeda et al. 2013); with an orbital period of just 8.5 h, it is sure to be a hellish world! I re-determine its mass using HARPS-N and HIRES observations in Chap. 4. Kepler-186f, was the first validated Earth-size planet to lie in the habitable zone of an M dwarf, where liquid water can be sustained (it is believed to be a key element for the emergence and survival of carbon-based life) (Quintana et al. 2014). The *Kepler* mission uncovered a great diversity of planets, which are giving us a unique insight on planet occurrence rates (see recent statistical studies by Howard et al. 2010; Mayor et al. 2011; Fressin et al. 2013; Petigura et al. 2013 and others) and shaping our theories of planet formation (see Lissauer et al. 2014a and references therein).

The initial aim of the *Kepler* mission was to find and characterise "Earth twins", i.e. rocky planets orbiting Sun-like stars in the habitable zone. Achieving this goal has been more difficult than anticipated, however, mainly because the intrinsic photometric variability of stars due to oscillations, granulation, spots, flares, etc. had been underestimated. Since the failure of two of its reaction wheels in May 2013, *Kepler* has been recycled into K2, which points to fields near the ecliptic plane for about 80 days at a time (Haas et al. 2014). Its photometric performance is still excellent, and it is already discovering transiting planets (Barclay 2014).

RV follow-up of *Kepler* candidates and future instruments In order to confirm the brightest and most exciting *Kepler* candidates, a replica of HARPS for the Northern hemisphere was designed as the *Kepler* field is not visible from the South. HARPS-N is mounted on the 3.57 m Telescopio Nazionale Galileo (TNG) at La Palma, Spain (Pepe 2010; Cosentino et al. 2012). It has now been in operation for three years and routinely achieves a precision better than $1 \text{ m} \cdot \text{s}^{-1}$. HARPS-N has already enabled the characterisation of several *Kepler* systems, including Kepler-78 (Pepe et al. 2013), Kepler-10 (Dumusque et al. 2014), two close-in giant planet hosts KOI-200 and 889 (Hébrard et al. 2014) and a close-in super-Earth host Kepler-93 (Dressing et al. 2015).

Several other high-precision spectrographs are currently being commissioned (see Pepe et al. 2014a and references therein). *Minerva* (Hogstrom et al. 2013) is due to start operations this year and will contribute to the follow-up of transiting planets found by K2 and TESS, the Transiting Exoplanet Survey Satellite (Ricker et al. 2015), to be launched in 2017. In the same year, CHEOPS (CHaracterising ExOPlanets Satellite, (Fortier et al. 2014) will be sent out to determine the radii of planets found in current RV surveys and for more precise photometric follow-up of targets identified by K2 and TESS.

A new generation of near-infrared spectrographs is also emerging, with CARMENES (Quirrenbach et al. 2013), which is expected to begin work this year, and SPiROU (Delfosse et al. 2013), scheduled for 2017. These spectrographs will be ideally suited to look for planets in the habitable zones of M dwarfs, which are more luminous in the infrared than in the visible. The effects of stellar activity, which are a major obstacle in RV searches, are less marked in this region of the spectrum.

The ESPRESSO spectrograph, to be mounted on the 8-m VLT at Paranal Observatory, Chile in 2016 is expected to achieve a precision of $0.1 \text{ m} \cdot \text{s}^{-1}$ (Pepe et al. 2014b). PLATO (Rauer et al. 2014), a photometry mission planned for launch in 2024, is a wide-field instrument like WASP, which will enable the discovery of Earth-radius planets in the habitable zones of their host stars. It will target bright stars, enabling much more precise determination of the planetary masses – if we can overcome the challenges imposed by stellar activity (see later paragraph). PLATO will perform asteroseismology on the host stars, enabling accurate determination of stellar parameters and ages. It will also allow us to explore the architecture of planetary systems as a whole, which will provide unique insights on planet formation.

High-precision RV measurements The key to measuring the RV of a star with high precision is to obtain a spectrum with as many lines as possible. High-resolution spectrographs are fitted with a grating, which splits the light into many wavelength orders, with the same resolution at all wavelengths; each order is then cross-dispersed by a grism in order to separate the different spectral orders spatially. The resultant spectrum is projected onto a high quality square CCD unit, as pictured in Fig. 1.4. Such a spectrum contains thousands of absorption spectral lines. We can create "line masks" of the strongest lines expected in the spectrum of a given star (eg. F, G, K) based on wavelength atlases of line positions measured in laboratory experiments,

Fig. 1.4 (Pre-) first light spectrum obtained by HARPS-N. The *horizontal* bands are the wavelength orders split by a grating, each of which is then split by a grism. This setup produces thousands of spectral lines, from which a precise RV measurement can be extracted. Picture captured by Francesco Pepe with a DSLR camera

and assuming the profile of the lines (e.g., Gaussian). We cross-correlate our observed spectrum with a line mask in order to determine the wavelength shift and mean shape of each line. From this procedure, we can create a "mean" line known as the cross-correlation function (CCF). It is centred at λ_{obs} and its shape is a combination of all the lines in the spectrum; the more lines we have, the better defined the CCF. The cross-correlation technique is commonly used in spectroscopic data reduction pipelines (eg. for HARPS and HARPS-N, see Baranne et al. 1996 and Lovis and Pepe 2007).

HARPS-N is a twin of HARPS, its most notable differences being that it is fed by an octagonal fibre, which scrambles the light more effectively. It also uses a Thorium-Argon lamp for the wavelength calibration, but experiments are being carried out with a laser comb which is currently being used to map the locations of the individual pixels on the CCD, to combat systematic errors caused by irregular pixel sizes. The RV uncertainty due to photon noise on HARPS and HARPS-N measurements can be reduced down to $0.5 \text{ m} \cdot \text{s}^{-1}$ with appropriate exposure times. The level of instrumental noise, arising mostly from wavelength calibration, is now of the order of a fraction of a $\text{m} \cdot \text{s}^{-1}$ (Mayor and Udry 2008; Dumusque et al. 2010).

Further descriptions of telescope and spectrograph setups as well as wavelength calibration methods (iodine cell, Thorium-Argon lamp, laser frequency comb), along with references for further information are given by Perryman (2011, pp. 16–21).

This thesis: towards breaking the stellar activity barrier We have now reached a level of precision where the most significant source of noise comes from the star itself. Observations have shown that activity-induced RV variations are of the order of $0.5 \text{ m} \cdot \text{s}^{-1}$ for a quiet dwarf star (Makarov et al. 2009); they can reach tens of $\text{m} \cdot \text{s}^{-1}$ in active stars, and in some cases up to $50 \text{ m} \cdot \text{s}^{-1}$ (Saar and Donahue 1997). In comparison, a super-Earth orbiting a Sun-like star at 0.1 AU induces a signal with an amplitude of just $1.4 \text{ m} \cdot \text{s}^{-1}$. A habitable Earth-mass planet has an RV signature of under $10 \text{ cm} \cdot \text{s}^{-1}$ (see Table 1.1). Activity-induced signals can therefore conceal and even mimic planetary orbits in RV surveys, and this has resulted in several false detections (Queloz et al. 2001a; Bonfils et al. 2007; Huélamo et al. 2008; Boisse et al. 2009, 2011; Gregory 2011; Haywood et al. 2014; Santos et al. 2014; Robertson et al. 2014 and others). I review the magnetic activity processes and features responsible for RV variability in Chap. 2.

To this day, various activity decorrelation methods have been tested (including the methods of Queloz et al. 2009; Hatzes et al. 2011; Aigrain et al. 2012; Haywood et al. 2014) but no simple and all-inclusive recipe has yet been proposed. During my thesis, I developed a new analysis technique to account for activity-induced signals in RV searches using Gaussian processes. I present my method in Chap. 3, and apply it to RV observations of CoRoT-7, Kepler-78 and Kepler-10 in Chap. 4.

Understanding the effects of stellar activity on RV observations is crucial to develop the next generation of more sophisticated activity models, and further improve our ability to detect and characterise low-mass planets. In Chap. 5, I explore the physical origin of stellar RV variability and identify a new activity proxy through the study of the Sun. It is the only star whose surface can be directly resolved at high resolution, and therefore constitutes an excellent test case.

References

Aigrain S, Pont F, Zucker S (2012) Mon Not R Astron Soc 419:3147
Auvergne M et al (2009) Astron Astrophys 506:411
Baglin A, Team C (1998) Proc Int Astron Union 185:301
Bakos G, Noyes RW, Kovács G, Stanek KZ, Sasselov DD, Domsa I (2004) PASP 116:266
Baranne A et al (1996) Astron Astrophys Suppl Ser 119:373
Barclay T (2014) American astronomical society meeting abstracts #224, vol 224
Batalha NM et al (2011) Astrophys J 729:27
Boisse I et al (2009) Astron Astrophys 495:959
Boisse I, Bouchy F, Hébrard G, Bonfils X, Santos N, Vauclair S (2011) Astron Astrophys 528:A4
Bonfils X et al (2007) Astron Astrophys 474:293
Borucki WJ et al (2011) Astrophys J 728:117
Cameron AC, Pollacco D, Hellier C, West R, the WASP Consortium and the SOPHIE and CORALIE Planet-Search Teams (2009) In: Proceedings of the international astronomical union, vol 4, pp 29
Campante TL (2015) Submitted to IST Press, pre-print. arXiv:1503.06113
Campbell B, Walker GAH (1985) Stellar radial velocities, vol 1, pp 5
Carter JA et al (2012) Science 337:556
Chaplin WJ et al (2014) Astrophys J Suppl Ser 210:1
Cosentino R et al (2012) Society of photo-optical instrumentation engineers (SPIE) conference series, vol 8446, pp 84461V
de Bruijne JHJ (2012) Astrophys Space Sci 341:31
Delfosse X et al (2013) SF2A-2013: Proceedings of the annual meeting of the French society of astronomy and astrophysics, pp 497, arXiv:1310.2991
Doyle LR et al (2011) Science 333:1602
Dressing CD et al (2015) Astrophys J 800:135
Dumusque X, Udry S, Lovis C, Santos NC, Monteiro MJPFG (2010) Astron Astrophys 525:A140
Dumusque X et al (2014) Astrophys J 789:154
Fortier A, Beck T, Benz W, Broeg C, Cessa V, Ehrenreich D, Thomas N (2014) Society of photo-optical instrumentation engineers (SPIE) conference series, vol 9143, pp 91432J
Fressin F et al (2011) Astrophys J Suppl Ser 197:5
Fressin F et al (2013) ApJ 766:81 arXiv:1301.0842
Gregory PC (2011) MNRAS 415:2523 arXiv:1101.0800
Haas MR et al (2014) American astronomical society meeting abstracts #223, vol 223
Hatzes AP et al (2011) Astrophys J 743:75
Haywood RD et al (2014) Mon Not R Astron Soc 443(3):2517–2531
Hatzes AP (2014) Nature 513:353
Hébrard G et al (2014) Astron Astrophys 572:93
Hilditch RW (2001) An introduction to close binary stars. Cambridge University Press, Cambridge
Hogstrom K et al (2013) American astronomical society meeting abstracts #221, vol 221
Holman MJ et al (2010) Science 330:51
Howard AW et al (2010) Science 330:653
Huélamo N et al (2008) Astron Astrophys 489:L9
Kjeldsen H, Christensen-Dalsgaard J, Handberg R, Brown TM, Gilliland RL, Borucki WJ, Koch D (2010) Astronomische Nachrichten 331:966
Koch DG et al (2010) Astrophys J 713:L79

Latham DW, Stefanik RP, Mazeh T, Torres G (1989) Bull Am Astron Soc 21:1224
Léger A et al (2009) Astron Astrophys 506:287
Lindegren L (2009) American astronomical society, IAU symposium #261. Relativity in fundamental astronomy: dynamics, reference frames, and data analysis 27 April–1 May 2009 Virginia Beach, VA, USA, #16.01; Bull Am Astron Soc 41:890, 261, 1601
Lissauer JJ et al (2011a) Nature 470:53
Lissauer JJ et al (2011b) Astrophys J Suppl Ser 197:8
Lissauer JJ, Dawson RI, Tremaine S (2014a) Advances in exoplanet science from Kepler. Nature 513:336–344
Lissauer JJ et al (2014b) Astrophysical Journal 784:44
Lissauer JJ et al (2013) Astrophys J 770:131
Lovis C, Pepe F (2007) Astron Astrophys 468:1115
Makarov VV, Beichman CA, Catanzarite JH, Fischer DA, Lebreton J, Malbet F, Shao M (2009) Astrophys J 707:L73
Marcy GW, Butler RP (1994) Cool stars, stellar systems, and the sun, vol 64, pp 587
Matthews JM et al (2000) Cool stars, stellar systems, and the sun, vol 203, pp 74
Mayor M, Queloz D (1995) Nature 378:355
Mayor M, Udry S (2008) Phys Scripta T130:014010
Mayor M et al (2003) The Messenger 114:20
Mayor M et al (2011) Astron Astrophys, pre-print arXiv:1109.2497
Mayor M, Lovis C, Santos NC (2014) Nature 513:328
Murdin P (2002) Encyclopedia of astronomy and astrophysics, vol 1
Orosz JA et al (2012a) Astrophys J 758:87
Orosz JA et al (2012b) Science 337:1511
Pepe F (2010) The HARPS-N project, Technical report
Pepe F et al (2013) Nature 503:377
Pepe F, Ehrenreich D, Meyer MR (2014a) Nature 513:358
Pepe F et al (2014b) Astronomische Nachrichten 335:8
Perruchot S et al (2008) Society of photo-optical instrumentation engineers (SPIE) conference series, vol 7014, pp 70140J
Perryman M (2011) The exoplanet handbook by Michael Perryman
Petigura EA, Marcy GW, Howard AW (2013) Astrophys J 770:69
Pollacco DL et al (2006) Publ Astron Soc Pac 118:1407
Queloz D et al (2000) Astron Astrophys 354:99
Queloz D et al (2001a) Astron Astrophys 379:279
Queloz D et al (2001b) The Messenger 105:1
Queloz D et al (2009) Astron Astrophys 506:303
Queloz D et al (2010) Astron Astrophys 517:L1
Quintana EV et al (2014) Science 344:277
Quirrenbach A et al (2013) EPJ Web Conf 47:05006
Rauer H et al (2014) Exp Astron 38:249
Reuyl D, Holmberg E (1943) Astrophys J 97:41
Ricker GR et al (2015) J Astron Telesc Instrum Syst 1:4003
Robertson P, Mahadevan S, Endl M, Roy A (2014) Science 345:440
Rowe JF et al (2014) Astrophys J 784:45
Saar SH, Donahue RA (1997) Astrophys J 485:319
Sanchis-Ojeda R, Rappaport S, Winn JN, Levine A, Kotson MC, Latham DW, Buchhave LA (2013) Astrophys J 774:54
Santos NC et al (2014) Astron Astrophys 566:35
Seager S, Mallen Ornelas G (2003) Astrophys J 585:1038
Sozzetti A (2011) EAS Publ Ser 45:273
Strand KA (1943) Publ Astron Soc Pac 55:29
Struve O (1952) The Observatory 72:199

References

Torres G et al (2011) Astrophys J 727:24
van de Kamp P (1969) Astron J 74:757
Vogt SS et al (1994) Society of photo-optical instrumentation engineers (SPIE) conference series, vol 2198, pp 362
Wolszczan A, Frail DA (1992) Nature 355:145 (ISSN 0028-0836)

Chapter 2
Stellar Activity as a Source of Radial-Velocity Variability

The key to breaking the activity barrier in exoplanet detections lies in our understanding of the physical origin and temporal structure of stellar RV variability. This chapter provides a review of the manifestations of magnetic activity, their impact on photometric and spectroscopic observations, and the analysis techniques that have been developed in recent years to account for activity-induced RV signals. I also present the target selection criteria I proposed to pick "magnetically manageable" stars for HARPS-N RV follow-up.

The signatures of magnetic activity span a wide range of spatial and temporal scales. We tend to naturally think of things in terms of their physical sizes. But when it comes to looking at stars, we see them as minuscule point-like objects in the night sky, and it is impossible to observe their surfaces with high resolution (except for the Sun). We can thus only gather limited information about spatial structures on the stellar surfaces.

The time-dependent nature of observations allows us to watch the surfaces of stars change and evolve over time and thus form a detailed picture of the various time scales at play. The temporal structure of the signals we observe can tell us a lot about the stars they originate from. In the first part of this chapter, I will present each stellar activity timescale, starting from acoustic oscillations that evolve within minutes, up to magnetic cycles that last decades.

2.1 Magnetic Activity and Its Manifestations

Before we look at the temporal and spatial diversity of magnetic activity signatures, however, let us catch a glimpse at how magnetic fields are produced within stars. It has long been found that a dynamo process operates within the stellar interior (Babcock 1961; Parker 1963, see Tobias 2002 and references therein). Over recent years, helioseismic studies and sophisticated numerical models have taught us much about the Sun's internal dynamics, although much debate remains (see review by Charbonneau 2010). Current theories of solar dynamo processes are detailed in Choudhuri (2007).

The Sun, like all stars, generates energy in its core through the process of nuclear fusion. This energy is carried outwards in the form of radiation through the radiative zone, where photons undergo a random walk which takes about 10 million years (Lockwood 2005). The radiative zone constitutes the majority of the Sun's interior. Above it lies a convective zone which takes up around 30% of the interior (in radius). In less massive stars, the radiative layer is thinner or may not be present at all—for stars with masses less than $0.35\,M_\odot$ (corresponding to early M spectral type), convection is the dominant mechanism for energy transport throughout the star (Hansen and Kawaler 1994; Chabrier and Baraffe 1997). Hot fluid cells are driven upwards due to buoyancy forces. The cells cool when they reach the stellar surface and eventually sink back, and so on.

In Sun-like stars, the radiative and convective layers are separated from each other by a thin layer called the tachocline (Spiegel and Zahn 1992; Miesch 2005). In this region, strong radial shearing forces arise due to the transition between the uniformly rotating radiative zone and the differentially rotating convection zone. It is now generally accepted that this shear is the source of the stellar magnetic dynamo, which is responsible for stellar activity (Tobias 2002). Fully convective stars, in which there is no tachocline, have a different type of dynamo which can result in both basic magnetic field topologies (Morin et al. 2008) or very complex ones (Chabrier and Küker 2006). Fully radiative stars have very weak and unordered fields, if any, and it is unclear how they are created (see Walder et al. 2012 and references therein). A small subset, however, such as the chemically peculiar A stars, have very strong magnetic fields, but these appear to be fossil fields rather than dynamo-generated. Their configurations do not change with time (see Aurière et al. 2014 and references therein).

Light escapes from stars in the bottom layer of the stellar atmosphere: the photosphere. Above the photosphere lies the chromosphere, which is surrounded by the corona, which extends out into space through the solar wind. The photosphere is commonly regarded as the stellar surface and is peppered with granulation, spots and faculae: these are some of the signatures of stellar magnetic activity.

Useful textbooks and reviews on the topics covered in this chapter include Rutten and Schrijver (1994)—proceedings of *Solar Surface Magnetism*,[1] Schrijver and Zwaan (2000)—a book on solar and stellar magnetic activity, Hall (2008)—a review on chromospheric activity, and Reiners (2012)—a review on observations of magnetic fields in Sun-like stars. A great place to find general reviews and articles on solar and stellar activity is the *Living Reviews in Solar Physics*.[2] For more general information on stellar interiors and atmospheres, two classic textbooks are Novotny

---

[1] NATO Advanced Research Workshop, held in the Netherlands in 1993.
[2] Published by the Max-Planck-Institut für Sonnensystemforschung, Germany. Available online at: http://solarphysics.livingreviews.org/.

(1971) and Gray (1992). I also provide more specific references throughout this chapter.

2.1.1 Minutes: Oscillations

Stars breathe. Their internal pressure constantly fluctuates by tiny amounts; this creates acoustic waves going through the star's interior, which result in the formation of ripples on the stellar surface. These waves were first observed on the Sun by Leighton et al. (1962); later on, Bedding et al. (2001) reported on the first clear detection of similar oscillations in a star other than the Sun, α Cen A. These oscillations, commonly known as p-modes, repeat on timescales of about 5–15 min and produce RV oscillations with an amplitude of a few m · s^{-1}.

To illustrate this, I retrieved RV observations of the bright Sun-like star μ Arae, taken at a 2-min cadence over an 8-night run for an asteroseismic study (Bouchy et al. 2005). The RV variations recorded on one of the nights are plotted in Fig. 2.1. The p-mode oscillations are clearly visible, particularly in panel (c), which shows a close-up on an oscillation with a period of about 8 min. We can confirm the presence of this signal by looking at the Lomb–Scargle periodogram (Lomb 1976; Scargle 1982; Zechmeister and Kürster 2009; see Sect. 2.3.2.1) of the dataset, displayed in Fig. 2.2. We also see other peaks at 5 and 11 min, which arise from p-mode oscillations that have a slightly different frequency.

We can average out the RV effects of these short frequency oscillations simply by making sure our observations are at least 10 min long (Dumusque et al. 2010). It is common practice with HARPS and HARPS-N to make 15-min observations in order to cancel their effect.

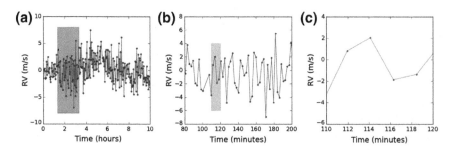

Fig. 2.1 RV observations of the bright star μ Arae (also known as HD 160691, $V = 5.1$ mag), monitored at high-cadence (100-s exposures with 31-s of dead time in between) as part of an 8-night HARPS run in June 2004 (ESO program 073.D-0578, Bouchy et al. 2005). Panel **a** shows observations made over one night. Panel **b** is a zoom-in over a 2 h period, and panel **c** is a zoom-in over 10 min. Panel **a** clearly shows the 2-h granulation signal, while panels **b** and **c** highlight the p-modes

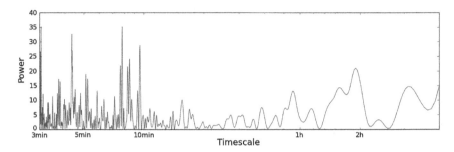

Fig. 2.2 Lomb–Scargle periodogram of the RVs of μ Arae plotted in Fig. 2.1. The strong peaks close to 4 and 8 min are caused by the p-modes, while the peak at about 2 h is due to granulation motions

2.1.2 Minutes: Flares and Coronal Mass Ejections

The magnetic energy stored in active regions and their surroundings can lead to sudden releases in the form of large eruptions known as flares or coronal mass ejections (see Hathaway 2010 and references therein). These events lead to sudden and sharp increases of brightness, and have been observed on other Sun-like stars in *Kepler* lightcurves (see Walkowicz et al. 2011 and others).

These dramatic events are rare in the sort of low-activity stars suitable for planetary RV searches, and easily identified in RV observations as they will generate spikes of several tens of m · s^{-1} in the mean RV variations of a star, and show strong signatures in the Hα emission profile (Reiners 2009).

2.1.3 Minutes to Hours: Granulation

A small patch of the Sun's surface is pictured in Fig. 2.3, revealing the bright and dark granulation structures in impressive detail. This pattern originates from the convective motions taking place below the surface: hot fluid cells rise up to the surface, forming large and bright patches, which sink once they have cooled and become dense enough for gravity to pull them back down.

Granules have a diameter of a few hundred kilometres, and a lifetime of the order of about 8 min (Bahng and Schwarzschild 1961; Hall 2008). There also exist larger structures called mesogranules; these have lifetimes of 30–40 min (Roudier et al. 1998). The largest convective cells, known as supergranules, have sizes of order 40–50 Mm and remain on the stellar surface for about a day (Del Moro et al. 2004).

This short frequency bustling leads to variations in brightness, which we can now easily probe on Sun-like stars thanks to high precision, short cadence *Kepler* photometry (Gilliland et al. 2011).

2.1 Magnetic Activity and Its Manifestations

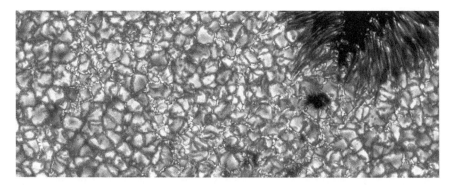

Fig. 2.3 Granulation on the solar surface, observed at high resolution with the Swedish 1-m telescope, at La Palma. A sunspot is visible in the *top-right corner*—its dark centre, known as the umbra, is surrounded by a lighter region known as the penumbra. Convection cells are visible on the photosphere surrounding the spot. Image credit: Vasco Henriques, http://www.isf.astro.su.se/gallery/images/2010/ (link valid as of March 2015)

The vertical motions of convection produce RV variations of the order of around $2\,\text{km} \cdot \text{s}^{-1}$. Since there are about 1 million granules on the visible hemisphere of the Sun at any time, the global RV variations can be thought of as arising from fluctuations in the number of granules present. The fluctuations obey Poisson statistics, so they are on the order of the square root of the number of granules, reducing the observed RV flicker on the granulation timescale from $2\,\text{km} \cdot \text{s}^{-1}$ down to $2\,\text{m} \cdot \text{s}^{-1}$ (Lindegren and Dravins 2003). The first evidence of the such RV variations were observed by Labonte et al. (1981) and Kuhn (1983) on the Sun. Several years later, Kjeldsen et al. (1999) reported on the first clear evidence of periodic fluctuations due to granulation in a star other than the Sun, α Cen A.

We can go back to the RV observations of μ Arae, shown in Fig. 2.1 to identify its granulation signature. In panels (a) and (b), we can see variations over a longer timescale than that of the p-mode oscillations. An inspection of the periodogram in Fig. 2.2 reveals peaks close 1 and 2 h, which can be attributed to granulation (although in this particular case, these peaks may be aliases of a longer 8-h cycle that can be seen in plots of the full 8 nights of data, presented in Bouchy et al. 2005).

As for p-modes, adapting our observing strategy to mitigate their effect on RV measurements also works for granulation. Taking several RV measurements on each night observed (generally 2 to 3 measurements) spaced by about 2 h significantly reduces the RV effects of granulation (Dumusque et al. 2010).

2.1.4 Days and Longer: Gravitational Redshift

Photons escaping from the photosphere are slowed down by the strong gravitational potential of the star; the photons become redshifted, causing the centres of the spectral lines to shift. In the case of the Sun, this results in a shift in RV of order $600\,\text{m}\cdot\text{s}^{-1}$ (Lindegren and Dravins 2003). The magnitude of this shift depends on the stellar radius, and Cegla et al. (2012) calculated that a change of 0.01 % in the radius of the Sun would induce a shift of about $6\,\text{cm}\cdot\text{s}^{-1}$ in RV This is enough to mimic or mask the orbital reflex motion of the Earth in our solar system if the radius fluctuations are taking place on a long enough timescale. Cegla et al. (2012) found that fluctuations occurring over 10 days or longer became significant. This means that we do not have to worry about the effect of p-modes; however, changes in the granulation pattern on the stellar surface and the Wilson depression of starspots (see Solanki 2003 and references therein) can potentially produce radius fluctuations that would yield small, but significant RV variations (see Cegla et al. 2012 and references therein). Variable gravitational redshift is not a major source of activity-induced RV variations, however.

2.1.5 Stellar Rotation Period: Spots, Faculae and Plage Regions

Stellar surface features such as spots and networks of faculae induce photometric and spectroscopic variations that are modulated by the rotation period of the star. These signals pose a serious challenge to the detection of exoplanets. Various decorrelation methods have been developed (see Sect. 2.2), but no simple and all-inclusive recipe has yet been found.

2.1.5.1 Sunspots and Starspots

Sunspots are seen as dark areas on the surface of the Sun (see top right corner of Fig. 2.3). Hale (1908) was the first to notice Zeeman splitting of lines produced in these dark regions and deduced that sunspots are regions of strong magnetic fields. They are indeed areas where magnetic flux loops emerge from the solar surface (Solanki 2002); the magnetic fields inhibit part of the outgoing convective heat flux, resulting in areas of reduced brightness and temperature. The spots usually appear as pairs of opposite magnetic polarity. For a detailed review of the general properties of sunspots, refer to Solanki (2003).

Observations of similar dark and magnetic structures on the surfaces of other stars have led to the concept of stellar spots. Starspots are defined as "an environment in which magneto-convective interaction significantly suppresses convective energy transport over an area large enough that a structure forms that is cool and dark relative to the surrounding photosphere" (Schrijver 2002). They are similar to sunspots in many aspects; the most notable difference is that starspots can attain huge sizes and can exist near the poles of their stars (Strassmeier 2009). In the next few paragraphs, I briefly describe the main properties of sunspots and starspots.

2.1 Magnetic Activity and Its Manifestations

Many of the papers cited below are part of the Proceedings of the First Potsdam Thinkshop on Sunspots and Starspots (Strassmeier et al. 2002). I also learned much on starspots from reviews by Berdyugina (2005) and Strassmeier (2009). Thomas and Weiss (2008) is a comprehensive book on starspots and sunspots. These are all excellent sources of information to find out more about the physical properties of sunspots and starspots.

Sunspots have temperatures ranging from 600 to 1800 K less than the surrounding photosphere, and starspots have similar temperature differences ranging from 500 to 2000 K (Schrijver 2002). Since spots have lower temperatures than the rest of the stellar surface, they appear darker. We can determine their magnetic field strength and the magnetic filling factor over the whole stellar surface via Zeeman splitting of spectral lines, using high-resolution spectra; we are not yet able, however, to disentangle these two quantities (Saar 1991; see Reiners 2012 and references therein).

As spots grow and decay, they induce variations in photometry that are modulated by the star's rotation (see some example lightcurves in Fig. 2.11). As a star rotates, one half of the disc is moving towards us, while the other half is moving away; as a result, the flux emitted by the approaching half is blueshifted, while the receding half is redshifted. If the stellar surface presents no features, the Doppler shifts from both sides cancel each other out and the spectral line profile is undisturbed, as pictured in the left diagram of Fig. 2.4. A starspot coming in and out of view as the star rotates, as shown in the subsequent diagrams of Fig. 2.4, blocks some of the flux of the star, inducing an imbalance between the redshifted and blueshifted halves of the star. This produces an asymmetry in the shape of the total line profile, thus shifting its centroid by a small amount. These perturbations to the line profile translate into RV variations of the order of $1\,\mathrm{m}\cdot\mathrm{s}^{-1}$ for sunspots (Lagrange et al. 2011; Makarov et al. 2009);

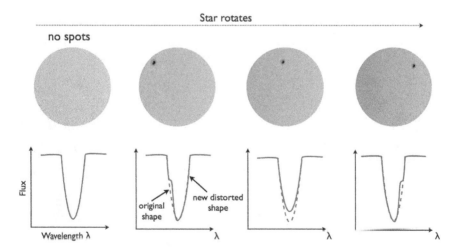

Fig. 2.4 Diagram illustrating how flux blocked by starspots on the rotating stellar disc induces asymmetries in the spectral lines, leading to variations in RV

starspot-induced RV variations can be much greater for more active, more rapidly rotating stars.

We can monitor these line-profile distortions to track the evolution of spots. This technique, commonly known as *Doppler imaging*, was first applied to the rapidly-rotating star HR 1099 by Vogt and Penrod (1983) to reconstruct a stellar surface brightness map. It was later applied to the rapidly-rotating K dwarf AB Doradus, by Donati and Collier Cameron (1997) to map the stellar magnetic flux distribution. The procedure is illustrated in Fig. 2.5, with HARPS cross-correlations functions (CCFs) of sunlight scattered from the bright asteroid Vesta (I present a detailed analysis of these observations in Chap. 5). The CCFs obtained from each observation are first stacked on top of one another to obtain a time series of line profiles, as shown in panel (a); I then compute the mean line profile shown in panel (b); finally, I subtract this mean profile from each CCF of the time series, in order to reveal asymmetries in the line profiles, as shown in panel (c). These distortions are produced by sunspot groups drifting across the solar disc. Using this technique, we can deduce the latitude of the spot groups and therefore construct maps of the stellar surface. The Doppler imaging technique works best on (fast-rotating) stars with long-lived spot groups that will remain on the stellar disc for several rotations.

Sunspots have sizes ranging from 1,500 to 20,000 km, and even the largest spots will only cover a small fraction of the solar surface (<1 %). The average sunspot coverage on the Sun is typically between 0.0001 and 0.1 %, depending on the phase of the solar cycle (Strassmeier 2009). It is trickier to determine the sizes of starspots are we cannot resolve their surfaces at high resolution. The amplitude of photometric variations depends on the size of a spot, or group of spots present at a given longitude, but it also depends on the contrast in brightness of the spot group, which itself depends on the temperature contrast. This is a complex issue, and many studies have been carried out to disentangle these two quantities using Doppler imaging (Catalano et al. 2002 and references therein). Doppler imaging uncovered starspots of all sizes

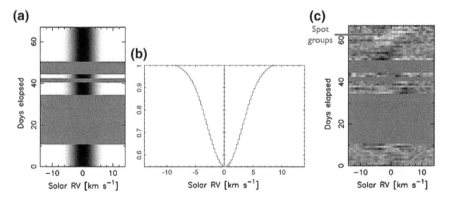

Fig. 2.5 Doppler-imaging the Sun! in three simple steps: **a** make a time series of the CCFs; **b** compute the mean line profile; **c** subtract the mean line profile from the time series to reveal line-profile distortions caused by sunspots and groups of faculae trailing across the solar disc

ranging from 0.1 up to 22 % of the stellar surface (Strassmeier 2009). There have also been observations of huge polar spots on some stars (see Schrijver 2002; Strassmeier 2009 and references therein).

Sunspots as well as small starspots live from a few days up to several weeks (Schrijver 2002; Allen 1973; Hussain 2002). In general, the lifetime of a spot is proportional to its size (Berdyugina 2005); spots decay by diffusing out into the surrounding photosphere, so spots with a relatively larger area-to-perimeter ratio should take more time to disappear (Solanki 2003; Petrovay and van Driel-Gesztelyi 1997; Robinson and Boice 1982). This has been confirmed observationally via *Kepler* data (see Helen Giles, MSci project at University of St Andrews, results to be published).

On the Sun, sunspots are always found between the latitudes of ±35°; they migrate closer to the equator as the solar cycle progresses (see Sect. 2.1.6). A similar behaviour is seen on other stars, although spots can also be found at much higher latitudes. Sunspots preferentially appear at so-called active longitudes, where increased magnetic activity in a localised region causes spots to manifest repeatedly in the same region (Berdyugina and Usoskin 2003). Active longitudes have also been observed on other stars (Olah et al. 1989; Lanza et al. 2009 and others). They rotate in phase with the stellar rotation (modulo differential rotation), and could explain a persistent coherent starspot signal. Ivanov and Kharshiladze (2013) found that prominent solar active longitudes can survive for up to 20 solar rotations.

2.1.5.2 Faculae and Plage

Faculae are small bright pores on the stellar photosphere and are associated with strong magnetic fields (Spruit 1976). On the Sun, they are around 100 K hotter than the rest of the photosphere (Thomas and Weiss 2008). They are found in the intergranular lanes, and surround spots—spots are always surrounded by faculae and plage. Faculae, however, can exist on their own and are grouped together into large networks. Because they are shaped as thin flux tubes with bright walls, they are best seen near the stellar limb. Faculae have lifetimes of a couple of hours (Hirayama 1978), but groups of faculae can remain on the stellar surface for several weeks and will last for several stellar cycles. Faculae always appear before spots and will also outlive them. As we will see in Sect. 2.1.6, old and slowly-rotating Sun-like stars are dominated by faculae over starspots.

Plage regions are bright areas of the chromosphere made up of small bright points known as flocculi (see Zirin 1966 and references therein). Flocculi, or facular bright points (Soltau 1993) are surrounded by thin and dark upward moving jets known as spicules (Roberts 1945; Zirin 1966). Similarly to faculae, flocculi (and spicules) have short lifetimes of 15 to 30 min and appear brighter close to the limb (although as we get too close to the edge they become obscured by the tall spicules).

Chromospheric plage regions map closely to faculae and spots in the underlying photosphere. Plages and faculae tend to be located near sunspots, although their rela-

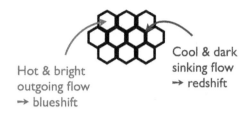

Fig. 2.6 Schematic representation of convection cells on the stellar surface

tionship is not yet understood (Hall 2008; Schrijver 2002). Athay (1974)[3] provides further in-depth discussions on the nature of plage regions and possible relations between photospheric and chromospheric active regions (see Bumba and Ambroz 1974 in particular).

Emission lines such as Ca II H&K, Hα and the Ca II triplet lines form at the level of the chromosphere, and are good indicators of plage regions (Mallik 1996; Cincunegui et al. 2007). Activity indicators based on the Ca II H&K lines are discussed further in Sect. 2.2.1.1.

The photometric effect of faculae is negligible as they are not significantly brighter than the quiet photosphere and they are evenly spread on the stellar disc; they do, however, induce a strong signature in spectroscopic observations. The strong magnetic fields present in faculae and spots act to inhibit the convection process taking place at the stellar surface. Let us think back on granulation, and take a closer look at its spatial structure. Granules can be approximated as bright hexagonal cells; they are surrounded by dark intergranular lanes, as illustrated in Fig. 2.6. The material in the intergranular lanes is cooler and therefore more compact than the hot fluid of the granules, which means that over the whole stellar disc, we see a larger proportion of hot, uprising fluid over cool, sinking material (Gray 1989). This results in a net blueshift, with a magnitude of about $200\,\text{m} \cdot \text{s}^{-1}$ on the Sun (see Meunier et al. 2010 and references therein). The presence of networks of faculae suppresses part of this blueshift.[4] As they evolve, active regions can lead to RV variations of up to $8\text{–}10\,\text{m} \cdot \text{s}^{-1}$ for the Sun (Meunier et al. 2010), as well as the active Sun-like star CoRoT-7 (Haywood et al. 2014, see Chap. 4). Suppression of convective blueshift is thought to play a dominant role in activity-induced RV variations on Sun-like stars, particularly in the case of faculae/plage, which are thought to cover a much larger fraction of the stellar surface than spots.

2.1.5.3 Other Possible Sources of Surface Velocity Fields

As I have shown in this chapter so far, the stellar photosphere and chromosphere are bustling with all kinds of constantly evolving and moving features such as granula-

---

[3]Chromospheric fine structure: proceedings from IAU Symposium no. 56 held at Surfer's Paradise, Qld., Australia, 3–7 September 1973.

[4]Starspots also act to suppress convection, but they contribute little flux and therefore do not play a significant role in this process (see Dumusque et al. 2014 and references therein).

tion, spots, networks of faculae and plage regions. There are other phenomena that may induce RV variations, such as ~50 m · s^{-1} horizontal inflows towards active regions recently found on the Sun (Gizon, Duvall and Larsen (2001); Gizon, Birch and Spruit (2010)). Such photospheric velocity fields may affect the RV curve (particularly when located towards the limb, as they are horizontal flows) even if they have no detectable photometric signature.

2.1.6 Decades: Magnetic Cycles

The Sun has an activity cycle of 11 years (Schwabe 1844; Hathaway 2010). Progression into the cycle towards higher activity is observed as an increase in the number of sunspots, faculae, plages and is also accompanied by a more frequent occurrence of violent events such as prominences and coronal mass ejections (see Hathaway 2010 for a detailed review). At minimum activity, sunspots are located at latitudes of 30–35°. As the cycle advances, they are found closer and closer to the equator. This results in a pattern known as the "butterfly diagram" (Maunder 1904).

In 1966, Dr Olin Wilson founded the HK Project, a survey of 1296 Sun-like stars within 50 pc of our Sun that was undertaken in an effort to characterise their activity levels and see whether other stars also displayed activity cycles similar to the Sun's (Wilson 1968). Observations were made with the Coudé scanner attached to the 100-inch telescope at the Mount Wilson observatory. The fluxes in the Ca II H and K lines were measured, as it was already known for the Sun that the flux in these lines is correlated with the number of sunspots, i.e. an indicator of activity (Leighton 1959; Sheeley 1967—see Sect. 2.2.1.1). Wilson (1978) presented results on 91 stars after the first 11 years of observations, showing the first evidence for cyclic stellar variability.

In 1977, an improved photoelectric spectrometer was built by Dr Arthur Vaughan and placed on the 60-in. telescope, also on Mount Wilson (Vaughan et al. 1978). The activity index S was developed by Vaughan et al. (1978) in order to quantify levels of activity; I will define it further in Sect. 2.2.1.1. Values of the S-index (or the $\log R'_{HK}$, also see Sect. 2.2.1.1) for over a thousand stars were calculated and reported in Duncan et al. (1991), Baliunas et al. (1995), Henry et al. (1996), and Lockwood et al. (1997).

Baliunas et al. (1998) noticed that the majority of stars surveyed showed periodic variations with cycles of at least 7 years, and some lasting more than 30 years. A quarter of the stars displayed variability but with no apparent periodicity, while the remaining 15 % seemed to show no activity at all. The survey ran until 2003, and to this date remains the most extensive survey on stellar activity and variability. A similar and complementary project at Lowell Observatory (Arizona, USA) with the Solar-Stellar Spectrograph was initiated in 1994 (Hall et al. 2007) to record activity in Sun-like stars, and has made more than 20,000 observations since.

In parallel, it was found that as Sun-like stars get older, they rotate more slowly and their magnetic activity levels decline (Wilson 1963; Kraft 1967; Skumanich 1972,

Noyes et al. 1984). This means that young stars tend to rotate faster and be more active, and old stars like the Sun rotate more slowly and have lower activity levels.

Further studies on the variability of Sun-like stars by Radick et al. (1998) and Lockwood et al. (2007), based on the Mount Wilson and Lowell stellar samples, revealed the existence of distinct types of variability patterns. In young stars, photometric variations tend to be anti-correlated with chromospheric variations ($\log R'_{HK}$), which indicates that their surfaces are dominated by spots during phases of high activity levels. In the case of older, slowly-rotating stars such as the Sun, photometric variations are positively correlated with chromospheric variations. This means that their surfaces are dominated by faculae rather than spots. The dividing line between these two types of variability was found to be at $\log R'_{HK} = -4.7$. The Sun, with $\log R'_{HK} = -4.96$ lies just below this limit and its surface is thus faculae-dominated.

The long-term, continuous observations obtained over the last decades have given us an invaluable insight into the time-variant activity patterns of stars other than our Sun. We are still left to wonder, however, about the spatial evolution of stellar activity over these long timescales. For example, do starspots migrate across the surface in the same way that sunspots do? Sanchis-Ojeda et al. (2011) and Sanchis-Ojeda and Winn (2011) showed that it is possible to deduce the latitude of starspots occulted by planetary transits. Llama et al. (2012) successfully recovered spot locations from transit occultations in the continuous, high-precision photometry provided by the *Kepler* satellite over its 3.5-year lifetime. They carried out simulations of magnetic cycles for a range of cycle durations and found that it is possible to track the migration of spots on active stars with short activity cycles; with a longer dataset, they would be able to characterise spot-belts on Sun-like stars. A couple of simulated "butterfly diagrams" with different activity levels are shown in Fig. 2.7.

Stellar magnetic cycles can produce significant RV variations, in some cases of up to $25 \, \text{m} \cdot \text{s}^{-1}$ (Lovis et al. 2011). So far, we have been searching mostly for short-period planets ($P < 50$ days). A few large RV programs with HARPS and HIRES have been running for 5–10 years (including Lovis et al. 2011; Marcy et al. 2014). They are only just beginning to catch glimpses of magnetic cycles in RV observations. In their recent detection of an Earth-mass planet with a 3-day orbit around α Centauri

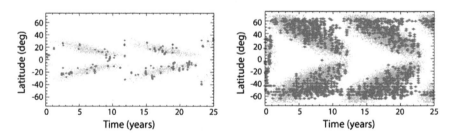

Fig. 2.7 Simulated "butterfly diagrams" for stars with an 11-year activity cycle like the Sun, with low and high activity levels, respectively. The *blue dots* show spots that have been recovered through bumps in the transit lightcurve, while the small *grey dots* represent the input butterfly pattern, for reference. These plots were made by Dr Joe Llama, based on work from Llama et al. (2012)

2.1 Magnetic Activity and Its Manifestations   25

B, from the analysis of over 3 years of data, Dumusque et al. (2012) found that the long-term activity-induced RV variations followed the variations in $\log R'_{HK}$. They were therefore able to model the RV variations assuming a linear relationship with $\log R'_{HK}$. This may not work as well if the planet's orbital period is comparable to the magnetic cycle, however, and as we begin to look for Jupiters and Saturns with orbital periods comparable to magnetic activity cycle durations, this will become a growing concern.

2.1.7 Timescales: Summary

The surface of a star is constantly bustling with magnetic activity, which leads to a plethora of RV perturbations. On the shortest timescales (oscillations, granulation), we can average out most of the effects on RV by adapting our observing strategy. On timescales of the order of decades, assuming a linear relationship between long-term activity RV variations and $\log R'_{HK}$ variations will work as a first approximation, although as we begin to look for long-period planets we are going to require more effective methods and proxies.

The most complex activity-induced RV variations, which cause the most trouble in today's RV exoplanet surveys arise from processes taking place on the stellar rotation timescale. Strongly magnetised photospheric features such as starspots and networks of faculae (as well as chromospheric plage regions) inhibit convective motions occuring just below the stellar surface, thus suppressing part of the blueshift naturally resulting from granulation. This effect can lead to variations in RV of up to $10\,\mathrm{m\cdot s^{-1}}$ (Meunier et al. 2010; Haywood et al. 2014). In addition, starspots coming in and out of view as the star rotates induce an imbalance between the redshifted and blueshifted halves of the star which translates into an RV modulation of the order of $1\,\mathrm{m\cdot s^{-1}}$ (Lagrange et al. 2011; Makarov et al. 2009). There may even be other processes at play which induce significant RV variations (Haywood et al. 2014), such as horizontal flows toward active regions (Gizon et al. 2001, 2010) or other unknown processes, whose impact on RV variations will require further investigation.

Identifying informative and reliable proxies for activity-driven RV variations has become crucial for exoplanet detection and characterisation. In the next section, I outline the various proxies and activity decorrelation techniques that have been developed for RV planet searches so far.

2.2 Existing Treatments for Activity-Induced RV Variations

This section provides a detailed summary of the analysis techniques developed to identify planetary signals in the presence of stellar activity. The methods of harmonic decomposition, pre-whitening and nightly offsets were initially developed to determine the mass of transiting super-Earth CoRoT-7b, so if you wish to place them in a more "historical" context, you can read the introduction on CoRoT-7 in Chap. 4, Sect. 4.1.1 in parallel.

2.2.1 Spectroscopic Activity Indicators

The following indicators, derived from the same stellar spectra used to measure the stellar RV, are affected by stellar activity only, so any variations present in RV observations but not seen in these indicators may point to a planetary signal.

2.2.1.1 Activity Indicators Based on Ca II H & K Line Fluxes

The S-index was first used by Vaughan et al. (1978). In his review on stellar chromospheric activity, Hall (2008) defines it as "a dimensionless ratio of the emission in the line cores [of Ca II H & K] to that in two nearby continuum bandpasses on either side of the H and K lines". The S-index can therefore be expressed as:

$$S = \alpha \frac{\Psi_H + \Psi_K}{\Psi_V + \Psi_R}, \qquad (2.1)$$

where Ψ_H and Ψ_K refer to the fluxes in the cores of the H and K lines respectively, and Ψ_V and Ψ_R refer to the fluxes in the bands on the violet and red sides of the H and K lines. The term α is a normalisation factor. The amount of flux measured in the reference passes, however, depends on spectral type so the S index cannot be used to compare stars of different colours. The S-index also varies when applied to measurements taken with different instruments, since the level of transmission of the bandpasses depends intrinsically on the instrumentation used. Middelkoop (1982) was the first to apply a correction term to the S-index in order to overcome its color dependence.

The R'_{HK} index was introduced by Noyes et al. (1984) in an effort to propose an activity index independent on spectral type and instrument design. Hall (2008) defines it as "the fraction of a star's bolometric luminosity radiated as chromospheric H and K emission". This is expressed in mathematical terms as (Martínez-Arnáiz et al. 2010):

$$R'_{HK} = \frac{\Psi'_H + \Psi'_K}{\sigma T_{\text{eff}}^4}, \qquad (2.2)$$

where σ denotes the Stefan–Boltzmann constant and T_{eff} is the effective temperature of the star. The primes on the fluxes Ψ are to show that the chromospheric contribution of the reference star has been subtracted. Note that there Ψ' values in this context are also in the form of fluxes measured at the stellar surface, rather than those received by the observer, to be consistent with the use of σT_{eff}^4. The R'_{HK} index is widely used, usually in logarithmic units.

2.2.1.2 Indicators Derived from the Cross-Correlation Function

As I described in the previous chapter, in order to measure the RV of a star all the lines of a spectrum are combined together to produce a mean line profile known as the cross-correlation function (CCF). Its shape reflects the shape of all the lines in the spectrum, which are affected by physical processes taking place in the stellar atmosphere, where these lines form. Here I present two measures of the shape of the CCF, that have been used in previous studies to identify activity-induced signals in RV data.

Full width at half-maximum (FWHM) The full width at half-maximum of the CCF, or FWHM is shown in Fig. 2.8. The FWHM is determined by the stellar rotation rate, i.e. the $v \sin i$ of the star (Desort et al. 2007). Since younger, fast rotating stars tend to be more active, it ensues that the FWHM gives a general indication of the levels of magnetic activity of a star. The FWHM also incorporates the intrinsic width of the line due to thermal and turbulent motions in the stellar photosphere.

The FWHM changes as a spot of facular region crosses the stellar disc, in order to conserve the area enclosed by the line profile (see Fig. 2.4). RV perturbations arising from the flux blocked by starspots on a rotating star are therefore correlated with variations in the FWHM. This indicator has been used by a number of studies, including Queloz et al. (2009), Hatzes et al. (2010) and Lanza et al. (2010) in the case of CoRoT-7 (see Chap. 4, Sect. 4.1.1) to identify activity-related signals.

Bisector of the cross-correlation function (BIS) A more sophisticated measure is the bisector of the CCF (see Fig. 2.9). It is defined as a measure of the general asymmetry of the lines of a spectrum (Voigt 1956), and was first used for exoplanet detection by Queloz et al. (2001). A more rigorous definition of the bisector, given by Perryman (2011) is: "the locus of median points midway between equal intensities on either side of a spectral line, thereby dividing it into two halves of equal equivalent width". For a line profile with a perfect Gaussian shape, this would be a straight

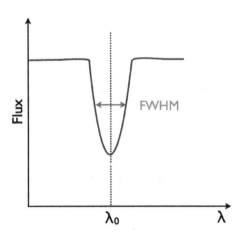

Fig. 2.8 The full width at half-maximum of the cross-correlation function

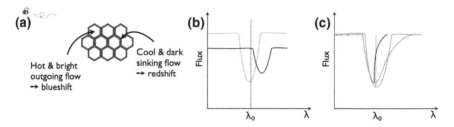

Fig. 2.9 How the shape of the line bisector is affected by surface granulation. *Panel* **a** Schematic representation of granulation pattern. *Panel* **b** Line profiles resulting from light emitted by the bright granular regions (*top, yellow line*) and dark intergranular regions (*bottom, brown line*). *Panel* **c** Effective line profile (*blue*), with its "C"-shaped line bisector (*red*); the undisturbed profile and its bisector are drawn in *dotted lines*. This figure was inspired from a similar figure in Dravins et al. (1981)

vertical line going through the middle of the line profile (dotted lines in Fig. 2.9c). However, the net blueshift produced by granulation on the stellar surface (explained back in Sect. 2.1.5.2) results in a bisector curved towards the top (see Fig. 2.9c). The granulation pattern is made of dark regions surrounding bright granules (panel (a); see Sect. 2.1.3). The bright upflowing granules produce the blueshifted line profile shown in yellow in panel (b), while the dark sinking intergranular flow leads to the redshifted line with a lower intensity, shown in brown on panel (b). The total line profile is the sum of these two profiles, as pictured in blue in panel (c). It is asymmetric, and its bisector (full red line) is curved at the top. Active regions that reduce this net blueshift will thus produce small distortions in the bisector. Many quantities have been defined in relation to the bisector, such as the bisector velocity span (Toner and Gray 1988), the curvature of the line bisector (Hatzes 1996), and the bisector inverse slope (Queloz et al. 2001); see Figueira et al. (2013) and references therein for more detail.

Desort et al. (2007) found that the FWHM, the BIS and photometric variations do not give enough information for slowly rotating, Sun-like stars (low $v \sin i$) to disentangle stellar activity signatures from the orbits of super-Earth-mass planets (see also Chap. 5, Section 5.4.2).

2.2.2 Nightly Offsets Method

This technique is very effective for short-period planets observed 2 to 3 times each night. It was successfully applied to the CoRoT-7 and Kepler-78 systems (Hatzes et al. 2010, 2011; Pepe et al. 2013; see Chap. 4).

Activity-related signals change on relatively long timescales (of the order of P_{rot}), whereas the planet's orbital period will be of a few hours (up to 1–2 days). In such a case, it is reasonable to assume that, on a given night, the rotation-modulated stellar activity contribution to the RV signal is roughly constant, and all the variations

2.2 Existing Treatments for Activity-Induced RV Variations

occuring over the span of a few hours are caused by the orbital reflex motion of the planet. This will work well for stars with low granulation "flicker" (see Sect. 2.3.2.3 which introduces the F8 statistic, a good measure of this noise source).

We can fit a linear function of the form:

$$m_i = A \cos(\omega t_i) + B \sin(\omega t_i) + C_j, \tag{2.3}$$

where A and B give the amplitude and phase of the RV signal, and there is an offset C_j for each night which represents the offset produced by the slowly varying activity modulation. The best-fit parameters can be determined via an optimal scaling procedure as follows. First, C_j is calculated for each night by taking the variance-weighted average of the data y_i in each single night:

$$\hat{C}_j = \frac{\sum_{i=1}^{n_j} y_{ij} w_{ij}}{\sum_{i=1}^{n_j} w_{ij}}. \tag{2.4}$$

The subscript j refers to each night and goes from 1 to the total number of nights, whereas the subscript i refers to each individual data point in each night and goes up to the number of points in each night (n_j). w_i are the inverse variance weights defined as:

$$w_i = \frac{1}{\sigma_i^2}, \tag{2.5}$$

where σ_i is the error associated with the data y_i.

The constant parameters A and B are found by performing the following summations over the whole dataset:

$$\hat{A} = \frac{\sum_{ij} [y_{ij} - \hat{C}_j - \hat{B} \sin(\omega t_{ij})] \cos(\omega t_{ij}) w_{ij}}{\sum_{ij} \cos^2(\omega t_{ij}) w_{ij}}, \tag{2.6}$$

and

$$\hat{B} = \frac{\sum_{ij} [y_{ij} - \hat{C}_j - \hat{A} \cos(\omega t_{ij})] \sin(\omega t_{ij}) w_{ij}}{\sum_{ij} \sin^2(\omega t_{ij}) w_{ij}}. \tag{2.7}$$

An iteration is then carried out until A and B both converge. For further detail on iterative optimal scaling, the reader may refer to Collier Cameron et al. (2006) or Keith Horne's *Ways of Our Errors*.[5]

---

[5]Unpublished as of August 2015 but available online at: http://star-www.st-and.ac.uk/~kdh1/ada/woe/woe.pdf.

2.2.3 Harmonic Decomposition

As shown by Jeffers et al. (2009), any starspot configuration can be modelled by a series of harmonics of $P_{\rm rot}$ containing only the first three or four Fourier terms.[6] Subtracting this model from the data will help reveal signals that do not originate from the star's activity. Harmonic decomposition is based on three parameters: the stellar rotation period, the number of harmonics and the coherence time. The rotation period can be determined via Lomb–Scargle or autocorrelation techniques, which I describe later in Sects. 2.3.2.1 and 2.3.2.2.

Harmonic decomposition can be implemented by fitting a Fourier series of the form:

$$m_i = m_0 + \sum_{k=1}^{l} [C_k \cos(k\,\omega t_i) + S_k \sin(k\,\omega t_i)], \qquad (2.8)$$

where the number of desired harmonics is given by l and m_0 is a constant. The best fit can be determined via an iterative optimal scaling procedure akin to that presented in Sect. 2.2.2.

In this case, the inverse variance weights are given by:

$$w_i = \frac{\mathcal{G}(t - t_i)}{\sigma_i^2}, \qquad (2.9)$$

where $\mathcal{G}$ is a Gaussian function defined as:

$$\mathcal{G}(t - t_i) = \exp\left[-\frac{1}{2}\left(\frac{t - t_i}{\tau}\right)^2\right]. \qquad (2.10)$$

τ is the coherence time. It governs the time interval over which each data point at time t_i retains its importance. τ is normally chosen to be slightly less than the rotation period of the star, so that it is short enough to filter out the slow varying signals (due to activity—starspots usually have lifetimes of about one rotation period or longer), but not so much that it will destroy short period signals.

This technique was applied to CoRoT-7 using the first three harmonics (Queloz et al. 2009; Hatzes et al. 2010), and up to the first six harmonics (Ferraz-Mello et al. 2011). It was found that the activity signal can be reproduced nearly perfectly using only the first three (Queloz et al. 2009), since for higher harmonics the amplitude of the signal becomes negligible.

---

[6] As a side note, this also means that we cannot reconstruct a map of the stellar surface solely based on its photometric variations!

2.2.4 Pre-whitening

A Fourier analysis is carried out to find the strongest period in the signal, and a sinusoidal fit with this period is subtracted from the data. We repeat this until the noise level is reached. This is a quick way to uncover the strongest periods present in the signal and to compose a periodogram. It is analogous to the CLEAN method derived by Högbom (1974) and Roberts et al. (1987). See also Queloz et al. (2009) and Hatzes et al. (2010).

2.2.5 The FF' Method

Aigrain et al. (2012) found that RV variations induced by starspots are well reproduced by a model consisting of the product of the photometric flux F and its first time derivative F'. It is assumed that the spots are small and limb-darkening is ignored. Spots influence the stellar RV by suppressing the photospheric surface brightness at the local rotational Doppler shift of the spot. Also, in areas of high magnetic field such as faculae, which on the Sun are often associated with spot groups, the convective flow is inhibited, leading to an attenuation of the convective blueshift (see Sect. 2.1.5.2). This effect is thought to be the dominant contribution to the total RV signal in the Sun (Meunier, Desort and Lagrange 2010).

As shown in Fig. 2.10, the RV perturbation ΔRV_{rot} to the star's RV incurred by the presence of spots on the rotating photosphere varies with both the flux deficit of the spot (F) and the line-of-sight velocity; F varies with foreshortening, so it has a cos phase, while the line-of-sight velocity varies with a sin phase (so it is proportional to F'). As derived in Aigrain et al. (2012), the RV perturbation due to a spot crossing the disc can be expressed as follows:

$$\Delta RV_{rot}(t) = -\frac{\dot{\Psi}(t)}{\Psi_0}\left[1 - \frac{\Psi(t)}{\Psi_0}\right]\frac{R_\star}{f}, \qquad (2.11)$$

where $\Psi(t)$ is the observed stellar flux, Ψ_0 is the stellar flux for a non-spotted photosphere and $\dot{\Psi}(t)$ is the first time derivative of $\Psi(t)$. $R_\star$ is the stellar radius. The parameter f represents the drop in flux produced by a spot at the centre of the stellar disc, and can be approximated as:

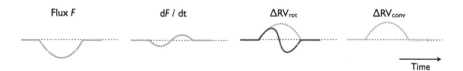

Fig. 2.10 The FF' method for a spot crossing the stellar disc. The RV variations induced by flux blocking (ΔRV_{rot}) and suppression of the convective blueshift (ΔRV_{conv}) are proportional to $F\,dF/dt$ and F^2, respectively

$$f \approx \frac{\Psi_0 - \Phi_{\min}}{\Psi_0}, \quad (2.12)$$

where $\Phi_{\min}$ is the minimum observed flux, i.e. the stellar flux at maximum spot visibility.

The effect of the suppression of convective blueshift on the star's RV produced by starspots and magnetised areas surrounding them, written as ΔRV_{conv}, is shown in Fig. 2.10. ΔRV_{conv} varies with foreshortening and the angle between the convective velocity vector and the line of sight. Both vary with the flux, so ΔRV_{conv} depends on F^2:

$$\Delta RV_{\text{conv}}(t) = \left[1 - \frac{\Psi(t)}{\Psi_0}\right]^2 \frac{\delta V_c \kappa}{f}, \quad (2.13)$$

where δV_c is the difference between the convective blueshift in the unspotted photosphere and that within the magnetised area, and κ is the ratio of this area to the spot surface (Aigrain et al. 2012). The two RV basis functions are pictured in Fig. 2.10.

This method does not depend on the period of rotation of the star, nor does it rely on complicated spot models. Aigrain et al. (2012) report on tests on HD 189733 that show it successfully reproduces previous results based on more complex models (Lanza et al. 2011). They also tested it on 600 *Kepler* targets and obtained promising results, given that the FF' method is very easy and quick to implement and could thus be applied to large datasets.

The FF' method is likely to provide an incomplete representation of activity-induced RV variations, however (Haywood et al. 2014—see Chap. 4, Sect. 4.1). The FF' method does not consider the broad-band photometric effect of faculae that are not physically associated with starspots; Aigrain et al. (2012) assume that their effect on ΔRV_{rot} is quite small as they tend to have low photometric contrast. Indeed, according to Lockwood et al. (2007), faculae become less important (relative to spots) in stars more active than the Sun (see Sect. 2.1.6). Faculae do, however, have a significant impact on the suppression of convective blueshift (Meunier et al. 2010); indeed we find that this effect dominates the total RV contribution induced by stellar activity (see Sect. 4.1.5.7). There are other phenomena that the FF' method does not account for, such as $\sim 50\,\text{ms}^{-1}$ inflows towards active regions recently found on the Sun (Gizon et al. 2001, 2010—see Sect. 2.1.5.3). Such photospheric velocity fields may affect the RV curve even if they have no detectable photometric signature. In addition, some longitudinal spot distributions have almost no photometric signature, so the FF' method would not account for them.

2.2.6 Existing Methods: Summary

The planet masses determined via all these methods should all agree, though the RMS scatter in the residuals may differ depending on how good the assumptions are. This reflects the fact that planet mass determinations are intimately tied to the methods

2.3 RV Target Selection Based on Photometric Variability

Since we do not yet know how to fully and reliably model all activity-induced RV variations, it is essential that we carefully pick stars for RV follow-up. Otherwise we may unknowingly choose a star with a rotation period that matches the orbital period of the planet, for example, and end up wasting huge amounts of telescope time for an imprecise, potentially even inaccurate planet mass determination.

Stars that have been observed by the high-precision photometry CoRoT and *Kepler* missions are ideally suited for potential RV follow-up, as we can learn a lot about their magnetic behaviour from their lightcurves.

The following question springs to mind: how do we define a magnetically "manageable" star for RV follow-up? This not only depends on the amplitude and frequency structure of activity variability; it is also tied with the mass and orbital period of the planet, and the decorrelation methods that we have developed to date.

For example, it is relatively easy to determine the mass of a super-Earth with a very short orbital period (typically less than 1 day), even if the host star is very active; the orbital and stellar rotation periods will be so different that we can take several observations per night and assume that all the variations produced within each night are solely due to the planet's orbital motion (see Sect. 2.2.2).

As a member of the Target Selection Tiger Team of the HARPS-N Guaranteed Time Observations collaboration, I was led to define magnetic manageability criteria to help us identify suitable *Kepler* candidates for HARPS-N RV follow-up (summer season of 2014).

2.3.1 Preliminary Target Selection Criteria

Before subjecting *Kepler* candidates to activity-related triage, a preliminary selection (from a pool of 600 available targets) was done by other members of the HARPS-N team. It includes the following criteria:

1. Target brightness: the majority of *Kepler* stars are too faint for ground-based RV follow up. This criterion was embedded in an estimation of the RV precision that would be achieved with the HARPS-N instrument. In this particular aspect, HARPS-N is a twin of HARPS, so we can use a relatively simple formula

determined by Bonfils et al. (2013). As a rule of thumb, stars should have a magnitude less than $V \approx 13$ mag.
2. Time required for a (3- or) 6-sigma detection: this can be calculated from the expected RV semi-amplitude of the planet candidate, which can itself be derived by assuming a bulk density (either a fixed value e.g. $\rho = 3 \text{ g} \cdot \text{cm}^{-3}$, or a radius-dependent value determined from mass-radius relations such as those derived by Weiss and Marcy 2014).
3. Number of planet candidates in the system; scientific interest related to each individual candidate (what is the scientific goal of this study: are we trying to populate the mass-radius diagram at a given radius/mass range? Are we trying to find other rocky mini-Neptunes like Kepler-10c?).
4. Observability: depending on how many months/years our survey is likely to last, determine the longest orbital periods it is reasonable to consider.
5. Asteroseismology information: this should be available if the target has been observed at short-cadence by *Kepler* or CoRoT. It will provide robust stellar parameters, which are essential to obtain a precise planet mass determination.
6. Previous follow-up: check the literature to see if this system has already been followed-up, and if this is the case, to assess whether additional RV measurements would be useful (eg., to determine the mass to a better precision, or to look for additional companions).

2.3.2 Generalised Lomb–Scargle Periodograms and Autocorrelation Functions

118 *Kepler* candidates survived these cuts. In order to identify activity selection criteria for these candidates, I computed Lomb–Scargle periodograms and autocorrelation functions (Lomb 1976; Scargle 1982; Edelson and Krolik 1988; Zechmeister and Kürster 2009) for the *Kepler* lightcurves of each star. I concatenated the lightcurves all the quarters together by dividing by the inverse variance weighted mean flux level for each quarter. This procedure is approximate but works well, as confirmed by visual inspection of a few lightcurves (see Fig. 2.11).

In the next two sections, I outline the concepts and main equations of the Lomb–Scargle periodogram and autocorrelation techniques.

2.3.2.1 Generalised Lomb–Scargle Periodogram

A natural first step as a planet hunter is to make a periodogram of the data, to get a first feel for what is in it. Most activity-induced signals will show some quasi-periodicity with a recurrence timescale related to the stellar rotation period, P_{rot} and/or its harmonics. The rotation of the star modulates both the lightcurve and RV curve.

2.3 RV Target Selection Based on Photometric Variability

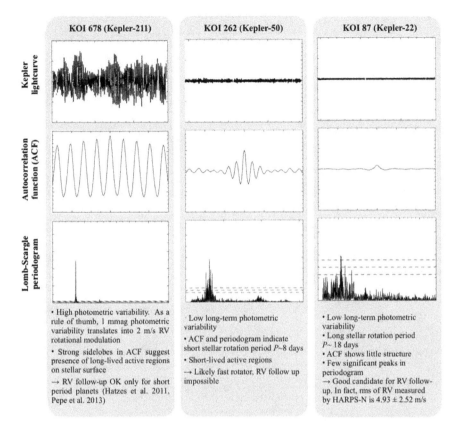

Fig. 2.11 Revealing the temporal behaviour of *Kepler* main sequence stars, through autocorrelation (time lag versus power) and Lomb–Scargle periodogram (frequency versus power) analyses of high-precision *Kepler* photometry, spanning all quarters of data. Note that all three lightcurve plots are on the same scale, ranging from −3 to +3 mmag. The plots were made using Andrew Collier Cameron's *dcfpgm.f* code

The following method is based on the techniques proposed by Lomb (1976), Scargle (1982) and Zechmeister and Kürster (2009). We can fit a sinusoid to our dataset:

$$m_i = A\cos(\omega t_i) + B\sin(\omega t_i) + C, \qquad (2.14)$$

where A and B are the amplitudes of the signal, C is an offset from zero, m_i is the fit to the photometric or RV data y_i at time t_i and $\omega = \frac{2\pi}{P_{\rm rot}}$ is the angular frequency associated with $P_{\rm rot}$.

The parameters A, B and C are calculated using iterative optimal scaling (also known as weighted least squares):

$$\hat{A} = \frac{\sum_i [y_i - \hat{C} - \hat{B}\sin(\omega t_i)]\cos(\omega t_i)\, w_i}{\sum_i \cos^2(\omega t_i)\, w_i}, \tag{2.15}$$

$$\hat{B} = \frac{\sum_i [y_i - \hat{C} - \hat{A}\cos(\omega t_i)]\sin(\omega t_i)\, w_i}{\sum_i \sin^2(\omega t_i)\, w_i}, \tag{2.16}$$

and

$$\hat{C} = \frac{\sum_i [y_i - \hat{B}\sin(\omega t_i) - \hat{A}\cos(\omega t_i)]\, w_i}{\sum_i w_i}. \tag{2.17}$$

The inverse-variance weights, w_i are defined in Eq. 2.5. The three parameters are summed over all data. The operation is repeated until a convergence threshold is met, for example when the change in each parameter is less than a given fraction of their associated uncertainty.

The right frequency is found by optimising the chi square (χ^2) on a grid of frequencies. The χ^2 is defined as:

$$\chi^2 = \sum_i \left[(y_i - m_i)^2\, w_i\right]. \tag{2.18}$$

The range of frequencies to be searched starts at the sidelobe frequency ($d\omega$) up to the Nyquist frequency (ω_{nyq}) at intervals given by the sidelobe frequency of the dataset. These two quantities are given by:

$$d\omega = \frac{2\pi}{t_{tot}} \tag{2.19}$$

and

$$\omega_{nyq} = \frac{n\pi}{t_{tot}}, \tag{2.20}$$

where n is the number of points in the dataset, and t_{tot} is the total span of the observations. Care should be taken to have both frequencies in the same units (rad · s^{-1} or deg · sec^{-1}).

Zechmeister and Kürster (2009) provide slightly different equations that are very easy to implement and quick to compute. We can also calculate false alarm probability levels for each signal. They are a measure of how likely it is for a signal with a given power to be caused purely by noise. Refer to Cumming (2004) and Collier Cameron et al. (2009) for a recipe on how to implement them.

2.3.2.2 Autocorrelation Function

Another way to determine the stellar rotation period is to compute the autocorrelation function of the data (Edelson and Krolik 1988). This technique provides us with

2.3 RV Target Selection Based on Photometric Variability

much more than just the period of the main signals in the data; an autocorrelation function is a star's activity identity card. The autocorrelation function conveys the same information as the Lomb–Scargle periodogram: with one glance at it we can tell the rotation period, whether the star has spots, how long they live for, how fast they decay, if there are many active regions, etc. The concept is the following: we take two copies of our dataset and shift them against each other by a small time interval τ at each step. The discrete autocorrelation of a dataset y (observation times t, uncertainties σ) is equal to (Edelson and Krolik 1988):

$$k_{i,j} = \frac{(y_i - \hat{y})(y_j - \hat{y})}{\sqrt{\sigma_i^2 \sigma_j^2}}, \qquad (2.21)$$

where $\hat{y}$ is the inverse-variance weighted average of the dataset, defined as:

$$\hat{y} = \frac{\sum_{i=1}^{n} y_i/\sigma_i^2}{\sum_{i=1}^{n} 1/\sigma_i^2}. \qquad (2.22)$$

The autocorrelation function can be normalised to 1 by dividing by its maximum. Each pair of points i, j is associated with the time lag:

$$\Delta t_{i,j} = t_j - t_i. \qquad (2.23)$$

For best results, the coherence length τ should be at least a few times longer than the spacing of the data, so that the autocorrelation looks smooth, but short-period signals longer than τ are still resolved.

For faster computation, only the positive (or negative) side of the autocorrelation function can be calculated and then simply mirrored onto the opposite side for plotting. The main recurrence timescale (in our case, the stellar rotation period) is the time lag of the centre of the first sidelobe of the autocorrelation function. A parabola can be fitted to this sidelobe in order to determine a more precise value, for example via iterative optimal scaling.[7]

2.3.2.3 Application to *Kepler* Lightcurves

Although I did implement my own versions of these techniques, for this target selection work I used Andrew Collier Cameron's code for discrete correlation functions & periodograms (dcfpgm.f), which I adapted for my own purposes. Figure 2.11 shows

---

[7]See Advanced Data Analysis course by Keith Horne online at http://star-www.st-and.ac.uk/~kdh1/ada/ada.html (link valid as of March 2015). The method of optimal scaling is explained in his notes from Lecture 5. His draft textbook *"The Ways of Our Errors"* is a gold mine to the astronomer looking for optimal data analysis methods.

lightcurves, Lomb–Scargle periodograms and autocorrelation functions for three example stars, with a short description of what we can learn about the behaviour of each star based on this information. In addition, I retrieved the following information:

- Two measures of the stellar rotation period—taken to be the strongest periodic signal identified in the Lomb–Scargle periodogram, and the time lag between the two main peaks of the autocorrelation function.
- The root mean scatter (RMS) of the *Kepler* lightcurves, obtained after removing points lying more than 5-sigma beyond average in order to remove any transits.
- The mean photometric error, σ_{av}, defined as the mean relative error in the *Kepler* photometry.
- The ratio between the amplitude of the main peak of the autocorrelation function (at zero time-lag) and the next highest peak: it tells us about the lifespans of active regions on the stellar surface.
- The F_8 statistic defined by Bastien et al. (2013), in units of parts per thousand (ppt). It corresponds to the RMS of the lightcurve over an 8-h timescale. It is computed by taking the RMS scatter in the photometry residuals after applying a boxcar filter of width 8 h. This scatter is caused by granulation and is known as the "8-h flicker". Faint stars will naturally display more variability in F_8, which we correct for by applying the relation used by Bastien et al. (2013), based on the *Kepler* magnitude V_{kepler} of the star (available in the *Kepler* input catalogue):

$$\log_{10} F_8 = -0.03910 - 0.67187 V_{kepler} + 0.06839 V_{kepler}^2 - 0.001755 V_{kepler}^3 \tag{2.24}$$

- $\log g$: I deduced this value from the F_8 statistic, using the relationship established by Bastien et al. (2013):

$$\log_{10} g = 1.15136 - 3.59637 \log_{10}(F_8) - 1.40002 \log_{10}(F_8)^2 - 0.22993 \log_{10}(F_8)^3 \tag{2.25}$$

It is useful to check the value of the $\log g$ as it gives an indication of whether the star is on the main sequence or if it is a giant or sub-giant star. Giants and sub-giants have lower surface gravity, bigger atmospheric scale heights, and hence fewer granules. The uncertainty in the number of granules on the stellar surface at any time is proportional to the square root of the number of granules (see Sect. 2.1.3), so the fractional uncertainty, and hence the noise (quantified in the F_8 statistic), goes up when there are fewer granules (Lindegren and Dravins 2003).

The plots in Fig. 2.12 illustrate the general behaviour of all 118 stars as a sample.

Stellar rotation periods The rotation periods obtained via both methods are plotted against each other in panel (a) of Fig. 2.12. They are in good agreement overall. In a few cases, especially at long periods, the period identified via Lomb–Scargle is twice as long as that of the autocorrelation. Period halving is a common problem at times when there are active regions on opposing hemispheres of the star. In such cases the autocorrelation sidelobes often alternate in amplitude, making it easier to

2.3 RV Target Selection Based on Photometric Variability

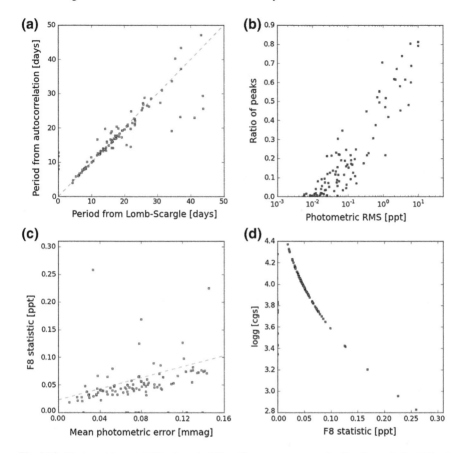

Fig. 2.12 Photometric variability characteristics of our stars as a sample. *Panel* **a** periods obtained through both methods plotted against one another; *panel* **b** Main-peak-to-first-sidelobe of autocorrelation function ratio, as a function of photometric RMS; *panel* **c** F_8 statistic plotted versus mean photometric error, σ_{av}; *panel* **d** $\log g$ values derived from the F_8 statistic. (The stars with zero Lomb–Scargle period or zero F_8 statistic are errors arising from the code, and further investigation would be required to solve this.)

identify the true period. I decided to use the stellar rotation period obtained from the autocorrelation method.

Care should be taken, however, mainly for two reasons:

- The majority of the stars in our sample show very little photometric variations; many stars display photometric RMS of order 0.1 or even 0.01 parts per thousand (ppt), as seen in panel (b) in Fig. 2.12. This means that it can be difficult to measure the rotation period reliably, even through the autocorrelation method which will exhibit weak sidelobe amplitudes.
- The lightcurves were reduced with *Kepler*'s PDC-MAP pipeline (Stumpe et al. 2012), which is a decorrelation method designed to remove patterns of instrumental

origin that are common to all stars in a given field of view of the *Kepler* CCD. It should not suppress signals of astrophysical origin, particularly variations due to the star's rotation and activity on short timescales. However, the *Kepler* Data Release 21 Notes caution that the PDC-MAP lightcurves should not be used to look for periodic signals longer than 20 days, as the pipeline erases long-term trends. On the other hand, McQuillan et al. (2014) and others have shown that it is possible to obtain reliable stellar rotational periods for a large number of *Kepler* stars.

Active region lifetimes The plot in panel (b) of Fig. 2.12 shows that there is a correlation between the amplitude of the star's photometric variations (RMS of the lightcurve) and the lifespan of active regions, which indicates that large active regions live longer. Starspots are thought to decay through diffusion, which takes place around the edge of the spots: larger spots, which have a smaller perimeter-to-area ratio will thus take longer to diffuse away (see references in Sect. 2.1.5.1).

2.3.3 Selection Criteria for "Magnetically Manageable" Stars

I settled on the following selection criteria:

1. Eliminate stars with a rotation period of less than 10 days, as we do not want to do RV follow-up on fast rotators (the cross-correlation profile would be very broad and yield a poorly-constrained RV measurement).
2. Require the RMS of the lightcurve to be less than 0.001 mag. This seemed like a reasonable threshold beyond which the star is showing a lot of modulation from starspots coming in and out of view.
3. In order to eliminate stars with anomalously high levels of granulation noise, I require:
$$\frac{(F_8 - 0.023)}{1000\sigma_{av}} < 0.5. \tag{2.26}$$

   The value 0.023 mag corresponds to the flicker "floor" seen in panel (c) of Fig. 2.12; an F_8 value below this limit indicates that the star is faint enough for the photon noise to dominate photometric variations induced by granulation. A high F_8 value makes it harder to average out the RV variations caused by granulation, even if we make a couple of observations separated by two hours. Indeed, most *Kepler* targets are so faint that we need to make 1800-s (30 min) exposures, and we therefore cannot really afford to take two per night.
4. Rotational and orbital timescales: it is more difficult to disentangle the orbit of a planet if its orbital period is close to the stellar rotation P_{rot} or its first harmonics. I discarded cases where the orbital period is within 2 days of P_{rot}, $P_{rot}/2$, and $P_{rot}/3$. I chose an interval of 2 days to be on the safe side since the stellar periods may not be very accurate.

5. It is also a problem if the orbital and stellar rotation timescales are too similar. I discarded cases where the rotation period was less than twice the orbital period.
6. However, if the timescales are *very* dissimilar, we can consider more active stars, i.e. a larger RMS. For these systems we can apply the nightly offset method detailed in Sect. 2.2.2. I therefore included targets for which the stellar rotational period is at least 10 times longer than the planet's orbital period, even if the photometric RMS was greater than 0.001 mag.
7. Distinguishing main sequence stars from giants: I require all viable candidates to have a log g greater than 3.5. This cutoff value is somewhat arbitrary, and it would be interesting to delve further into this to refine this criterion. The F_8 and log g are shown plotted against each other in panel (d) of Fig. 2.12.

2.4 Concluding Note: From Photometric to Radial-Velocity Variations

The next step would be to combine some of these indicators to predict the amplitude of activity-induced RV variations we might expect. The FF' method of Aigrain et al. (2012) (explained in Sect. 2.2.5) gives a recipe for doing exactly this. It does not fully account for the effect of faculae on the suppression of convective blueshift, or effects that have no photometric signature, however, so it is likely to largely underestimate the amplitude of activity-induced RV variations (see my work on CoRoT-7 in Chap. 4 and Haywood et al. 2014).

Based on RV data for CoRoT-7, Kepler-10 and Kepler-78 (see Chap. 4), as well as the findings of Aigrain et al. (2012), we can infer a simple rule of thumb: 1 mmag of photometric RMS results in $2\,\text{m}\cdot\text{s}^{-1}$ of activity-induced RV variations. Of course, the amplitudes are not the whole story; Bastien et al. (2014) found that the Fourier components of the lightcurve provide important clues about the complexity of the activity-induced RV variations. In this perspective, decoding the temporal structure of a star's lightcurve is a natural step towards understanding stellar RV variability.

The frequency structure of stellar signals reflects the character and personality of a star. We can use it to build a model for activity-induced RV variations, as I will show in the next chapter.

References

Aigrain S, Pont F, Zucker S (2012) Mon Not R Astron Soc 419:3147
Allen C (1973) Allen: astrophysical quantities, 3rd edn. The Athlone Press - Google Scholar (University of London)
Athay RG (1974) Proceedings of the International Astronomical Union, vol 56
Aurière M, Lignières F, Konstantinova-Antova R, Charbonnel C, Petit P, Tsvetkova S, Wade G (2014) In: Mathys G, Griffin ER, Kochukhov O, Monier R, Wahlgren GM (eds) Putting A stars into context: evolution, environment, and related stars, pp 444–450. arXiv:1310.6942

Babcock HW (1961) ApJ 133:572
Bahng J, Schwarzschild M (1961) Astrophys J 134:312
Baliunas SL, Donahue RA, Soon W, Henry GW (1998) Cool stars, stellar systems, and the sun, vol 154, pp 153
Baliunas SL et al (1995) Astrophys J 438:269
Bastien FA, Stassun KG, Basri G, Pepper J (2013) Nature 500:427
Bastien FA et al (2014) Astron J 147:29
Bedding TR et al (2001) Astrophys J 549:L105
Berdyugina SV (2005) Living Rev. Solar Phys. 2
Berdyugina SV, Usoskin IG (2003) Astron Astrophys 405:1121
Bonfils X et al (2013) Astron Astrophys 549:A109
Bouchy F, Bazot M, Santos NC, Vauclair S, Sosnowska D (2005) Astron Astrophys 440:609
Bumba V, Ambroz P (1974) Proc Int Astron Union 56:183
Catalano S, Biazzo K, Frasca A, Marilli E, Messina S, Rodonò M (2002) Astronomische Nachrichten 323:260
Cegla HM et al (2012) Mon Not R Astron Soc 421:L54
Chabrier G, Baraffe I (1997) Astron Astrophys 327:1039
Chabrier G, Küker M (2006) Astron Astrophys 446:1027
Charbonneau P (2010) Living Rev Solar Phys 7:3
Choudhuri AR (2007) In: Hasan SS, Banerjee D (eds) American institute of physics conference series. Kodai School on Solar Physics, vol. 919, pp 49–73
Cincunegui C, Díaz RF, Mauas PJD (2007) Astron Astrophys 469:309
Collier Cameron A et al (2009) Mon Not R Astron Soc 400:451
Collier Cameron A et al (2006) Mon Not R Astron Soc 373:799
Cumming A (2004) Mon Not R Astron Soc 354:1165
Del Moro D, Berrilli F, Duvall TLJ, Kosovichev AG (2004) Solar Phys 221:23
Desort M, Lagrange AM, Galland F, Udry S, Mayor M (2007) Astron Astrophys 473:983
Donati JF, Collier Cameron A (1997) Mon Not R Astron Soc 291:1
Dravins D, Lindegren L, Nordlund A (1981) Astron Astrophys 96:345
Dumusque X, Boisse I, Santos NC (2014) ApJ 796:132 arXiv:1409.3594
Dumusque X et al (2012) Nature
Dumusque X, Udry S, Lovis C, Santos NC, Monteiro MJPFG (2010) Astron Astrophys 525:A140
Duncan DK et al (1991) Astrophys J Suppl Ser 76:383
Edelson RA, Krolik JH (1988) Astrophys J 333:646
Ferraz-Mello S, Tadeu dos Santos M, Beaugé C, Michtchenko TA, Rodríguez A (2011) Astron Astrophys 531:A161
Figueira P, Santos NC, Pepe F, Lovis C, Nardetto N (2013) Astron Astrophys 557:93
Gilliland RL et al (2011) Astrophys J Suppl Ser 197:6
Gizon L, Birch AC, Spruit HC (2010) Annu Rev Astron Astrophys 48:289
Gizon L, Duvall TLJ, Larsen RM (2001) Proc Int Astron Union 203:189
Gray DF (1989) Astron Soc Pac 101:832
Gray DF (1992) Camb Astrophys Ser
Hale GE (1908) Astrophys J 28:315
Hall JC (2008) Living Rev Solar Phys 5:2
Hall JC, Lockwood GW, Skiff BA (2007) Astron J 133:862
Hansen CJ, Kawaler SD (1994) Stellar interiors. Physical principles
Hathaway DH (2010) Living Rev Solar Phys 7:1
Hatzes AP (1996) Publ Astron Soc Pac 108:839
Hatzes AP et al (2010) Astron Astrophys 520:A93
Hatzes AP (2011) Astrophys J 743:75
Haywood RD et al (2014) Mon Not R Astron Soc 443(3):2517–2531, 443, 2517
Henry TJ, Soderblom DR, Donahue RA, Baliunas SL (1996) Astron J 111:439
Hirayama T (1978) Publ Astron Soc Jpn 30:337

References

Högbom JA (1974) Astron Astrophys Suppl Ser 15:417
Hussain G (2002) Astronomische Nachrichten 323:349
Ivanov KG, Kharshiladze AF (2013) Geomagn Aeron 53:677
Jeffers SV, Keller CU, Stempels E (2009) In: Cool stars, stellar systems, and the sun (AIP), pp 664–667
Kjeldsen H, Bedding TR, Frandsen S, Dall TH (1999) Mon Not R Astron Soc 303:579
Kraft RP (1967) Astrophys J 150:551
Kuhn JR (1983) Astrophys J 264:689
Labonte BJ, Howard R, Gilman PA (1981) Astrophys J 250:796
Lagrange AM, Meunier N, Desort M, Malbet F (2011) Astron Astrophys 528:L9
Lanza AF et al (2009) Astron Astrophys 493:193
Lanza AF et al (2010) Astron Astrophys 520:A53
Lanza AF, Boisse I, Bouchy F, Bonomo AS, Moutou C (2011) Astron Astrophys 533:A44
Leighton RB (1959) Astrophys J 130:366
Leighton RB, Noyes RW, Simon GW (1962) Astrophys J 135:474
Lindegren L, Dravins D (2003) Astron Astrophys 401:1185
Llama J, Jardine M, Mackay DH, Fares R (2012) Mon Not R Astron Soc 422:L72
Lockwood GW, Skiff BA, Henry GW, Henry S, Radick RR, Baliunas SL, Donahue RA, Soon W (2007) Astrophys J Suppl Ser 171:260
Lockwood GW, Skiff BA, Radick RR (1997) Astrophys J 485:789
Lockwood M (2005) Solar outputs, their variations and their effects on Earth
Lomb NR (1976) Astrophys Space Sci 39:447
Lovis C et al (2011) Astron Astrophys, pre-print arXiv:1107.5325
Makarov VV, Beichman CA, Catanzarite JH, Fischer DA, Lebreton J, Malbet F, Shao M (2009) Astrophys J 707:L73
Mallik SV (1996) VizieR On-line Data Catalog 412:40359
Marcy GW et al (2014) Astrophys J Suppl Ser 210:20
Martínez-Arnáiz R, Maldonado J, Montes D, Eiroa C, Montesinos B (2010) Astron Astrophys 520:A79
Maunder EW (1904) Mon Not R Astron Soc 64:747
McQuillan A, Mazeh T, Aigrain S (2014) Astrophys J Suppl Ser 211:24
Meunier N, Desort M, Lagrange AM (2010) Astron Astrophys 512:A39
Middelkoop F (1982) Astron Astrophys 107:31
Miesch MS (2005) Living Rev Solar Phys 2:1
Morin J et al (2008) MNRAS 384:77 arXiv:0711.1418
Novotny E (1971) An introduction to stellar atmospheres and interiors. Oxford University Press, New York, 350 p, 1
Noyes RW, Weiss NO, Vaughan AH (1984) ApJ 287:769–773. doi:10.1086/162735, http://adsabs.harvard.edu/abs/1984ApJ...287..769N
Olah K, Panov KP, Pettersen BR, Valtaoja E, Valtaoja L (1989) Astron Astrophys 218:192
Parker EN (1963) ApJ 138:226
Pepe F et al (2013) Nature 503:377
Perryman M (2011) The exoplanet handbook by Michael Perryman
Petrovay K, van Driel-Gesztelyi L (1997) Solar Phys 176:249
Queloz D et al (2009) Astron Astrophys 506:303
Queloz D (2001) Astron Astrophys 379:279
Radick RR, Lockwood GW, Skiff BA, Baliunas SL (1998) Astrophys J Suppl Ser 118:239
Reiners A (2009) Astron Astrophys 498:853
Reiners A (2012) Living Rev Solar Phys 9
Roberts DH, Lehar J, Dreher JW (1987) Astron J 93:968
Roberts WO (1945) Astrophys J 101:136
Robinson RD, Boice DC (1982) Solar Phys 81:25
Roudier T, Malherbe JM, Vigneau J, Pfeiffer B (1998) Astron Astrophys 330:1136

Rutten RJ, Schrijver CJ (1994) NATO Adv Sci Inst (ASI) Ser C 433
Saar S (1991) IAU Colloq. 130: the sun and cool stars. Act, Magn, Dyn 380:389
Sanchis-Ojeda R, Winn JN (2011) Astrophys J 743:61
Sanchis-Ojeda R, Winn JN, Holman MJ, Carter JA, Osip DJ, Fuentes CI (2011) Astrophys J 733:127
Scargle JD (1982) Astrophys J 263:835
Schrijver CJ (2002) Astronomische Nachrichten 323:157
Schrijver CJ, Zwaan C (2000) Solar and stellar magnetic activity/Carolus J. Schrijver, Cornelius Zwaan. Cambridge astrophysics series, vol 34. Cambridge University Press, New York, 1
Schwabe M (1844) Astronomische Nachrichten 21:233
Sheeley NRJ (1967) Astrophys J 147:1106
Skumanich A (1972) Astrophys J 171:565
Solanki SK (2002) Astronomische Nachrichten 323:165
Solanki SK (2003) Astron Astrophys Rev 11:153
Soltau D (1993) In: The magnetic and velocity fields of solar active regions. Astronomical society of the Pacific conference series; Proceedings of the international astronomical union (IAU) colloquium no. 141, held in Beijing, China, 6–12 September 1992. Astronomical Society of the Pacific (ASP), San Francisco, pp 225
Spiegel EA, Zahn J-P (1992) Astron Astrophys 265:106
Spruit HC (1976) Solar Phys 50:269
Strassmeier KG (2009) Astron Astrophys Rev 17:251
Strassmeier KG, Washuettl A, Schwope A (2002) Astronomische Nachrichten 323:155
Stumpe MC et al (2012) Publ Astron Soc Pac 124:985
Thomas JH, Weiss NO (2008) Sustainability, 1
Tobias SM (2002) Philos Trans R Soc A: Math, Phys Eng Sci 360:2741
Toner CG, Gray DF (1988) Astrophys J 334:1008
Vaughan AH, Preston GW, Wilson OC (1978) Publ Astron Soc Pac 90:267
Vogt SS, Penrod GD (1983) Publ Astron Soc Pac 95:565
Voigt H-H (1956) Zeitschrift fur Astrophysik 40:157
Walder R, Folini D, Meynet G (2012) Space Sci. Rev. 166:145
Walkowicz LM et al (2011) Astron J 141:50
Weiss LM, Marcy GW (2014) Astrophys J 783:L6
Wilson OC (1963) Astrophys J 138:832
Wilson OC (1968) Astrophys J 153:221
Wilson OC (1978) Astrophys J 226:379
Zechmeister M, Kürster M (2009) Astron Astrophys 496:577
Zirin H (1966) A Blaisdell book in the pure and applied sciences. Blaisdell, Waltham, 1

Chapter 3
A Toolkit to Detect Planets Around Active Stars

In this chapter, I present a recipe to detect exoplanet orbits in RV observations in the presence of noise arising from stellar activity. I start by introducing Gaussian processes, which are a powerful and elegant way to model correlated noise. I will start from the very basics of Gaussian distributions, leading up to how I incorporate them in my model to account for stellar activity signals. I then present the model that I use to fit RV observations, and describe the Monte Carlo Markov Chain procedure that I apply to determine the best-fitting parameters of this model. Finally, I present the Bayesian model selection technique of Chib and Jeliazkov (2001) that I have implemented to estimate the most likely number of planets present in the system and therefore choose the most appropriate model.

3.1 Gaussian Processes

The first Sections (up to Sect. 3.1.1.5) of this introduction to Gaussian processes are based on a lecture given by Prof. David MacKay, from the Engineering Department at the University of Cambridge.[1] I am very thankful to him as it was only by watching his video that I finally understood what Gaussian processes really are! I thoroughly recommend watching it.

This chapter uses material from, and is based on, Haywood et al., 2014, MNRAS, 443, 2517.

[1] This lecture is posted online at http://videolectures.net/gpip06_mackay_gpb/ (link correct as of March 2015).

3.1.1 Definition

In statistical terms, a Gaussian process (GP) is defined as an n-dimensional random process, such that the joint probability distribution drawn from this process is a Gaussian distribution in n dimensions (Rasmussen and Williams 2006). Let us first look at a 1-dimensional Gaussian distribution.

3.1.1.1 1-Dimensional Gaussian Distribution

Any random process can be described by a probability distribution function. In the natural world, many processes follow a probability distribution function that is well described by a Gaussian "bell" shape (shown in Fig. 3.1). Imagine that we are on a field trip to measure the weights of blue tit birds. Before we catch a bird, we won't know exactly how much it weighs, but we can still take a pretty good guess because we know that the average weight of a tit is about 11 grams and that most blue tits weigh between 9 and 13 grams. Mathematically, we can write the probability P of measuring the blue tit's weight, y, as:

$$P(y) = \frac{1}{\sigma\sqrt{2\pi}} \exp^{-\frac{1}{2}\left(\frac{y-\mu}{\sigma}\right)^2}. \tag{3.1}$$

This is a Gaussian distribution centered at μ (=11 grams), with a standard deviation σ (=2 grams). Since we have not weighed the blue tit yet, this is a *prior* distribution. μ is the weight we are *most likely* to measure. Weighing birds, i.e. collecting data, will transform our prior beliefs into a posterior distribution.

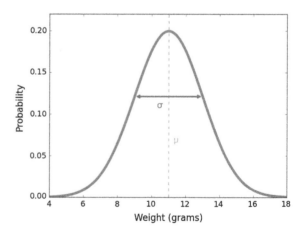

Fig. 3.1 1-dimensional Gaussian distribution, with mean μ and standard deviation σ

3.1.1.2 2-Dimensional Gaussian Distribution

Let's now imagine that we are also measuring the wing span of our blue tits. We expect it to be between 17 and 20 cm and to follow a bell-shaped distribution. We are now considering two variables, y_1 (weight) and y_2 (wing span). These two characteristics are intimately linked to each other, as bigger blue tits will weigh more and have a larger wing span: y_1 and y_2 are correlated with each other. This means that if we know the value of one, we will be able to take a better guess at the value of the other. Statistically, their probability distributions are joint together and form a 2-dimensional Gaussian distribution (which you can imagine as a hat shape). We can picture it as in Fig. 3.2.

The shape of the contours depends on how correlated the two variables are with each other; for example, they are weakly correlated in Fig. 3.2a, while they are strongly correlated in Fig. 3.2b. The quantitative relationship between these two variables (eg. $y_2 = 2y_1$ or $y_2 = y_1 + 5$) is not (yet) relevant. We shall leave this information encoded in a matrix **K**, which we treat as a "black box" for the time being.

In addition to wondering how precisely we can guess the wing span of a blue tit before we catch it, we can now ask: how much more precise will our guess of the wing span be after we have weighed it? Before we weigh the blue tit and thus measure y_1, the wing span y_2 will have a wide range of probable values. After we have measured y_1, the likely possibilities for y_2 narrow down to the range shown by the blue error bars in Fig. 3.2. If we now wish to predict y_2, knowing y_1, the values of y_2 will follow a Gaussian distribution centered at point μ_2. Note that if

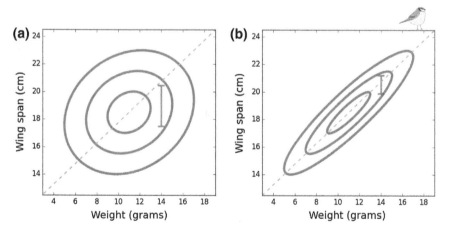

Fig. 3.2 2-dimensional Gaussian distributions, for two weakly (*panel a*) and strongly (*panel b*) correlated variables, with 1, 2 and 3-σ contours drawn in *red* (*Note* the error bars are not drawn to scale with respect to the bivariate distributions.)

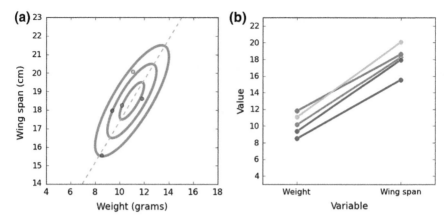

Fig. 3.3 The same 2-dimensional Gaussian distribution, displayed in the traditional (*panel a*) and new (*panel b*) representation

we had measured y_1 to be any other value, the most likely value of y_2, that is μ_2, would always fall on the straight line drawn on the plot. In other words, μ_2 is a linear function of y_1.[2]

As can be seen in Fig. 3.2b, in which the correlation between the two variables is stronger, the standard deviation of the distribution of y_2 after we have measured y_1 (which is now a *posterior* distribution) is much smaller than in the case of a weak correlation between the two parameters.

3.1.1.3 New Representation

In addition to the weight and wing span, we could measure as many features of our blue tits as we like, but plotting the joint distribution between more than 2 variables as in Fig. 3.2 would become tedious. Figure 3.3 illustrates how we can represent the same 2-dimensional Gaussian distribution of Fig. 3.2 in a different, simplified view. Imagine that we have now caught five birds (in statistical terms, we have drawn 5 samples from the joint prior distribution of the two variables "wing span" and "weight"). In the traditional view shown in panel (a), the set of measurements for each bird is marked with a dot; in the new representation in panel (b), it is represented by a line. In this new visualisation, the distributions of each variable y_1 and y_2 can be imagined standing vertically out of the page, centered at points μ_1 and μ_2.

---

[2] Remember this for later; it provides insight on the form of Eq. 3.16.

3.1.1.4 *n*-Dimensional Gaussian Distribution

In this new representation, picturing a Gaussian distribution with more than two dimensions becomes possible. An example of a joint 6-dimensional distribution is shown in Fig. 3.4. In panel (a), we draw several samples from the prior distribution; each of the coloured lines represents one sample, and tells us the value of each of the 6 variables. If we measure the value of one variable, such as in panel (b), this narrows the posterior distribution of the other variables, and so on. This only happens because the variables are all correlated with each other, and therefore depend on each other.

Each time we measure one of the variables of the process we are considering, we are making a cut through the n dimensional probability distribution space: this cut therefore has n-1 dimensions. This is analogous to taking a cut through a 3-dimensional sphere, and obtaining a 2-dimensional disc. If we measure all the

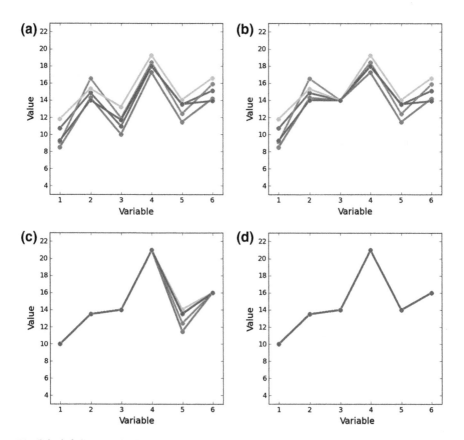

Fig. 3.4 A 6-dimensional Gaussian distribution. **a** we draw 5 samples from the prior distribution—no measurements have been made yet; **b** we measure variable 3; **c** we measure all the variables except variable 5; **d** all the variables have been measured

values of all the variables but one, we are left with a 1-dimensional Gaussian distribution which tells us about the behaviour of this last, unknown variable (panel (c)).

The nature of the correlation between y_i and y_j determines how little or how much the posterior distribution of y_j changes once we have measured y_i, and vice versa. Also, this correlation doesn't have to be the same between two other variables, say y_i and y_k. The correlation between each pair of variables of the process plays a key role in determining the process.

3.1.1.5 A Gaussian Process

Let's go back to Fig. 3.4, and use our imagination once more. If we just changed the labels on the axes, say the horizontal axis became "time", and the vertical axis became "flux", or "radial velocity"—the line in the plot suddenly looks like a fit to a set of observations! Yes, we can use a multi-dimensional Gaussian distribution to fit a dataset.

A Gaussian process is a non-parametric approach to fitting data. It is a Bayesian method. In the frequentist statistics approach, we start with a theory that we take for granted and we ask ourselves: *what is the probability that we will measure a given value?* We decide on the form of the model before we even start considering our observations, by specifying a parametric model, for example a sine function. Of course, such a model can, to some extent, be tailored to fit the observations; in the case of a sine function, we can determine the period, phase and amplitude that will provide the most optimal fit. The final model, however, will always be a sine curve and this may be a limitation in itself.

In the Bayesian world, we start with a dataset and use it to test our theory: *what is the likelihood that this model is correct?* Our model is non-parametric, which means that we do not impose the form of the model before we consider the observations; instead, we let the observations themselves determine the shape of the model. The only prior assumption that we make is about the way in which the data are correlated. We are making fewer prior assumptions and this gives us more freedom than having a model with a pre-determined shape.

We shall now ask: how do we define the correlations between points of a physical process? This is where the entity **K**, which I briefly mentioned back in the 2-dimensional case, comes into play!

3.1.2 Covariance Matrix K

For a dataset with n observations, **K** is an $n \times n$ matrix, which we refer to as the *covariance matrix*. Each element $\mathbf{K}_{i,j}$ gives the covariance between two dimensions y_i and y_j: this is a measure of how much these two variables change together.

3.1 Gaussian Processes

There are two options possible:

- The two variables are independent; if one changes, the other one does not. They are not correlated with each other. This is known as "white" noise.
- The two variables are dependent on each other; changing one will affect the other, because there is a correlation between them. This is commonly referred to as "red" noise.

3.1.2.1 Independent Data (White Noise)

Figure 3.5 a represents the covariance matrix of a set of 4 data points y_i, all independent from each other. Down the diagonal, we have $i = j$, so the entries correspond to the covariance of each point with itself; this is simply their variance, σ. In general, σ incorporates uncertainties induced by instrument systematics, weather conditions, etc.—it is just the error bar of the data. Because we are considering a case in which the observations are uncorrelated, all the non-diagonal elements are zero.

We can allow the variance to be different for each data point (σ_i), in which case the covariance matrix looks like in Fig. 3.5b.

Assuming a Gaussian distribution, the probability distribution P of a data point y_i can be written as:

$$P(y_i) = \frac{1}{\sigma_i \sqrt{2\pi}} \cdot \exp\left[-\frac{(y_i - \mu_i)^2}{2\sigma_i^2}\right]. \quad (3.2)$$

The joint probability distribution of all data points is:

$$P(\underline{y}) = \prod_{i=1}^{n} P(y_i), \quad (3.3)$$

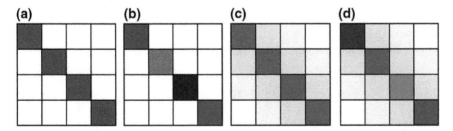

Fig. 3.5 Covariance matrix corresponding to: **a** *white* noise with a constant variance, **b** *white* noise with varying variance, **c** *red* noise and constant variance, **d** *red* noise and variable variance

and is referred to as the likelihood $\mathcal{L}$. Expanding this expression, we obtain:

$$\mathcal{L} = \left(\frac{1}{2\pi}\right)^{n/2} \cdot \left(\prod_{i=1}^{n} \frac{1}{\sigma_i}\right) \cdot \exp\left[-\sum_{i=1}^{n} \frac{(y_i - \mu_i)^2}{2\sigma_i^2}\right]. \quad (3.4)$$

We see that the sum in the exponential term corresponds to the chi-squared value (χ^2) of the data, so we have:

$$\mathcal{L} = \left(\frac{1}{2\pi}\right)^{n/2} \cdot \left(\prod_{i=1}^{n} \frac{1}{\sigma_i}\right) \cdot \exp\left(-\frac{\chi^2}{2}\right). \quad (3.5)$$

We obtain the best fit parameters by maximising this function, i.e. by minimising the χ^2.

3.1.2.2 Correlated Data (Red Noise)

We now consider data that are correlated with each other. The corresponding matrix is shown in Fig. 3.5c. The diagonal elements still represent the variance of the data, but some of the non-diagonal elements are now non-zero. The probability distribution function (i.e. the likelihood) is:

$$P(\underline{y}|\mathbf{K}) = \left(\frac{1}{2\pi}\right)^{n/2} \cdot \frac{1}{\sqrt{\det \mathbf{K}}} \cdot \exp\left[-\frac{1}{2}(\underline{y} - \underline{\mu})^T \mathbf{K}^{-1}(\underline{y} - \underline{\mu})\right], \quad (3.6)$$

where $\det \mathbf{K}$ is the determinant of $\mathbf{K}$. This is just a generalised expression of Eq. 3.4. In the case of white noise, the covariance matrix is equal to:

$$\mathbf{K} = \underline{\sigma}^2 \mathbf{I}, \quad (3.7)$$

where $\mathbf{I}$ is an $n \times n$ identity matrix. This reduces to a 1-dimensional array of size n, and hence leads to a simplified formulation of the likelihood that is proportional to the χ^2. The χ^2 can therefore only be used in the special case where the noise in our data is completely white.

Otherwise, we must use the generalised expression for $\mathcal{L}$:

$$\mathcal{L} = \left(\frac{1}{2\pi}\right)^{n/2} \cdot \frac{1}{\sqrt{\det \mathbf{K}}} \cdot \exp\left[-\frac{1}{2}(\underline{y} - \underline{\mu})^T \mathbf{K}^{-1}(\underline{y} - \underline{\mu})\right], \quad (3.8)$$

where the 2π term is a normalisation constant and the determinant of $\mathbf{K}$ acts as a penalty term for more complex models (Occam's razor).

We usually compute $\ln \mathcal{L}$:

$$\ln \mathcal{L} = -\frac{n}{2}\ln(2\pi) - \frac{1}{2}\ln(\det \mathbf{K}) - \frac{1}{2}(\underline{y} - \underline{\mu})^T \mathbf{K}^{-1}(\underline{y} - \underline{\mu}). \quad (3.9)$$

3.1.3 Covariance Function $k(t, t')$

Each element of the covariance matrix $\mathbf{K}$ is determined by a covariance function $k(t, t')$:

$$\mathbf{K}_{i,j} = k(t, t'), \tag{3.10}$$

where t and t' are associated with data points i and j.

Here are a few commonly used functions:

1. White noise:

$$k_1(t, t') = \theta_1^2 \cdot \delta_{t,t'}. \tag{3.11}$$

The term $\delta_{t,t'}$ is a Dirac delta function scaled according to the magnitude θ_1 of the white noise (usually given by σ_i). This is the simplest kind of covariance function. It leads to a diagonal matrix $\mathbf{K}$ and is almost always used, in combination with a more complex function.

2. Square exponential (Fig. 3.6a):

$$k_2(t, t') = \theta_1^2 \cdot \exp\left[-\frac{(t-t')^2}{\theta_2^2}\right]. \tag{3.12}$$

The hyperparameter θ_1 gives the maximum amplitude of the covariance between two points. The amplitude of the correlation falls exponentially over a (time)scale θ_2. This is the classic case in which we assume that points close to each other are more dependent on each other.

3. Periodic oscillation (Fig. 3.6b):

$$k_3(t, t') = \theta_1^2 \cdot \exp\left[-\sin^2\left(\frac{\pi(t-t')}{\theta_3}\right)\right]. \tag{3.13}$$

This kernel is ideal for a truly periodic, coherent signal, with a recurrence timescale θ_3.

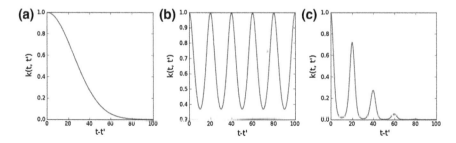

Fig. 3.6 Different types of covariance functions: **a** square exponential, **b** periodic, and **c** quasi-periodic. The hyperparemeters are: $\theta_1 = 1$, $\theta_2 = 35$, $\theta_3 = 20$, $\theta_4 = 0.5$

4. Quasi-periodic oscillation (Fig. 3.6c):

$$k_4(t, t') = \theta_1^2 \cdot \exp\left[-\frac{(t-t')^2}{\theta_2^2} - \frac{\sin^2\left(\frac{\pi(t-t')}{\theta_3}\right)}{\theta_4^2}\right]. \quad (3.14)$$

This kernel, of maximum amplitude θ_1, combines a square exponential term with a periodic variation at a fixed period θ_3. The quasi-periodicity evolves over a timescale θ_2. The hyperparameter θ_4 determines the amount of high frequency structure of the fit. The relative importance of the decay and periodicity terms in the exponential is dictated by the relative sizes of θ_2 and θ_4.

The parameters θ_j of the covariance function are known as the *hyperparameters* of the Gaussian process. In the classical statistics world, we fit data by determining the optimal values of the parameters of our (parametric) model, for example the period, phase and amplitude of a sine function; we find the best parameters in "data space". In the Bayesian world, when we are fitting data with a (non-parametric) (Gaussian) process, we find the optimal values of the hyperparameters of the covariance function; we determine the best parameters in "correlation space", or "covariance space". Doing so will give us much more freedom—we shall find out more as we go along!

The form of the covariance function is the main prior assumption we will make (the other priors being those imposed on the hyperparameters), so we need to think carefully about our choice—this is the subject of Sects. 3.1.4 and 3.1.5. We can also use Bayesian model selection tools to compare models with different covariance functions, in order to decide which one provides the best fit to the data, but is still the simplest function possible. For more detail on model selection, see Sect. 3.3.

3.1.4 Temporal Structure and Covariance

The shape of the covariance function is tightly linked to the temporal structure of the physical phenomenon that we are modelling. The perfect covariance function would look very similar to the autocorrelation function of the data.[3] This is illustrated in Fig. 3.7 for a star's lightcurve. The lightcurve displays strong variations due to starspots drifting across the stellar disc as the star rotates. Some starspots remain from one rotation to the next, making it easy identify the rotation period even in the lightcurve itself. The autocorrelation of the lightcurve shows a clear peak at the rotation period—it is not as strong as the peak at zero time lag, reflecting the evolution of spots over time. The autocorrelation does show some additional structure, but to a first approximation, it is well represented by a quasi-periodic covariance function (see Case 4 in Sect. 3.1.3).

---

[3] They cannot be completely identical since the covariance functions used with Gaussian processes are always positive definite, whereas the autocorrelation function oscillates about zero. They are very similar though (see Fig. 3.7), and it would be interesting to find out how they are related.

3.1 Gaussian Processes

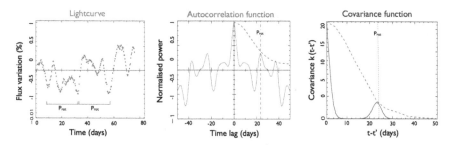

Fig. 3.7 *From left to right* the CoRoT-7 2012 *lightcurve*, its autocorrelation and the quasi-periodic covariance function used to fit the *lightcurve*

This intrinsic property of the covariance function sets the physical justification for using Gaussian processes to model activity-induced RV signals.

3.1.5 Gaussian Processes for Stellar Activity Signals

> [...] the 'jitter' formalism is limited, because it treats the activity signal as an independent, identically distributed Gaussian noise process.
> S. Aigrain et al. (2012)

Time dependent photometric and spectroscopic observations of stars tell us that activity-driven variations are not random or stochastic in nature. They follow a pattern, modulated by the star's rotation, which evolves according to the growth and decay of magnetically active regions on the stellar surface. The RV variations of a star are a tangled mess of activity and planetary signals, but in photometry, the activity and planetary signals are very distinct, and planet transits can easily be removed from the activity variations. As I have shown in Chap. 2, each star has its own unique behaviour. This temporal "character" is encoded in the periodogram or autocorrelation function of the star's lightcurve.

Due to their ability to memorise patterns of a given frequency structure, Gaussian processes are an ideal tool to model activity-induced variations. A quasi-periodic covariance function is an appropriate choice in this context. The evolution timescale corresponds to the average lifetime of starspots on the stellar photosphere, and the recurrence timescale is the stellar rotation period. We want the covariance function to go to zero for up to half of the stellar rotation cycle (assuming that the stellar rotation axis is inclined to 90° of the line of sight, which is the case for most transiting systems), to reflect the fact that we do not know what is happening to the surface features when they are facing away from us. Panel (c) of Fig. 3.6 shows clearly that the optimisation algorithm has selected a value of θ_4 that reduces the value of the covariance function to zero for roughly half of the stellar rotation cycle.

We can determine the hyperparameters of the covariance function using the star's off-transit lightcurve, since it shares the frequency structure of the star's magnetic activity. This is based on the assumption that the frequency structure of the covariance function representing the stellar activity should be the same for both the lightcurve and the RV curve. The Gaussian process fitting procedure is described in the following sections.

3.1.6 Determining the Hyperparameters θ_j

We can make rough guesses for the hyperparameters using our a priori knowledge of the phenomenon we are modelling, and this will provide a reasonable fit in most cases. However, it is best to let the data decide for themselves, so ideally we should use an optimisation method. In statistical jargon, determining the best hyperparameters to use is a procedure known as *training* the GP. I use a Monte Carlo Markov Chain (MCMC) in order to marginalise over all the hyperparameters—see a detailed description of this procedure in Sect. 3.2. The priors and ranges I normally apply to the hyperparameters are detailed in Table 3.1. We maximise the likelihood of Eq. 3.9 with respect to the hyperparameters.

Computationally, the matrix inversion required in this step means that this process is of order n^3. This means that it can get very slow for large datasets, and we may wish to consider binning the dataset we will be training the GP on (especially if it is a short cadence lightcurve!). In my code, I invoke a Cholesky decomposition (Press et al. 1986), which makes this equation very easy to implement. I explain how to fill the covariance matrix in the following Section.

3.1.7 Constructing the Covariance Matrix K

Once we have determined the covariance function and its hyperparameters, we can construct a covariance matrix for a dataset containing both red and white noise,

Table 3.1 Prior probability densities and ranges of the hyperparameters optimised via an MCMC procedure

Parameter	Symbol	Prior
Amplitude	θ_1	Modified Jeffreys (σ_{data})
Evolution timescale	θ_2	Jeffreys
Recurrence timescale	θ_3	Gaussian (P_{rot}, $\sigma_{P_{\text{rot}}}$) or Jeffreys (if P_{rot} not known)
Relative importance of evolution versus periodicity	θ_4	Uniform [0,1]

The knee of the modified Jeffreys prior is given in brackets. In the case of a Gaussian distribution, the terms within brackets represent the mean $\bar{x}$ and standard deviation σ. The terms within square brackets stand for the lower and upper limit of the specified distribution; if no interval is given, no limits are placed

3.1 Gaussian Processes

according to Eq. 3.10. In my model, I use a quasi-periodic covariance function (Case 4 of Sect. 3.1.3), with some additional white noise given by the error bars of the data:

$$k(t, t') = \theta_1^2 \cdot \exp\left[-\frac{(t-t')^2}{\theta_2^2} - \frac{\sin^2\left(\frac{\pi(t-t')}{\theta_3}\right)}{\theta_4^2}\right] + \sigma_i^2 \cdot \delta_{tt'}. \quad (3.15)$$

For each observation at time t, we calculate its distance in time to all the other observations of the dataset. In this way we can build a matrix whose values tell us about the degree of correlation between each point at time t and all other points at times t'.

3.1.8 Fitting Existing Data and Making Predictions

Gaussian processes are a machine learning tool: the covariance function is the "memory" of the GP, which learns from the data itself. The more data we have, the better we can determine the hyperparameters (see Sect. 3.1.6); by doing this, we are *conditioning* the GP. In turn, having better determined hyperparameters leads to a more probable fit to the data.

In this perspective, the covariance function not only allows us to determine the optimal fit to our observations; it also provides a means of interpolating the fit for times at which we do not have observations. In this case, we are asking how likely it is to measure a given value at a given time. If we make an observation at time t_i, what is the range of possibilities at time t_j?

We already know the covariance matrix **K** of the n existing data points $\underline{y} = (y_1, y_2, ..., y_n)$, as it is governed by the covariance function that we have already chosen and whose hyperparameters we have already determined. It has a size $n \times n$. These are the *training* points of the GP.

We define the covariance matrix $\mathbf{K_{pp}}$ for the m *test points* at times $\underline{t}_p = (t_{p1}, t_{p2}, ..., t_{pm})$ at which we want to predict the data. This matrix is populated with the same covariance function, and has dimensions $m \times m$.

The covariance matrix $\mathbf{K_p}$ of the *cross-terms* has to be evaluated too, as it will be used to calculate the errors on $\underline{y}_p$. It has dimensions $m \times n$.

The predicted data $\underline{y}_p$ are given by the mean of the predictive distribution, which is calculated as follows:

$$\underline{y}_p = \mathbf{K_p}.\mathbf{K}^{-1} \cdot \underline{y}. \quad (3.16)$$

If we are simply determining a fit to our data (i.e. we are not "predicting" data at new observation times), then $m = n$ so $\mathbf{K_p}$ is of size $n \times n$, and $\mathbf{K_{pp}} = \mathbf{K}$. Note that $\underline{y}_p$ is indeed a linear function of $\underline{y}$!

The errors associated with $\underline{y}_p$ are found by calculating the covariance of the predictive distribution, and then taking the square root of the diagonal elements of this matrix:

$$\underline{\sigma}_{y_p} = \sqrt{\mathrm{diag}[\mathbf{K_{pp}} - \mathbf{K_p} \cdot \mathbf{K}^{-1} \cdot \mathbf{K_p}^T]} \qquad (3.17)$$

We can interpret this last equation as taking the covariance matrix of the predicted times and "removing" the parts where the predicted times and measured times overlap: at these points, there is less uncertainty so the error is smaller.

Computational trick If you have already determined your hyperparameters, speed up your code by predicting your GP for one point at a time only; this way, $\mathbf{K_p}$ becomes a 1-dimensional vector, and $\mathbf{K_{pp}}$ reduces down to a scalar. Now you only need to solve one linear equation instead of doing a full Cholesky decomposition, which can save a lot of CPU time! Create a little subroutine that does this and simply call it m times.

3.1.9 A Word of Caution

It is good to remember that samples from the predictive distribution don't behave like the mean of the predictive distribution: the error bars $\underline{\sigma}_{y_p}$ are just as important as the predicted data $\underline{y}_p$ themselves. The choice of covariance function is crucial—it is important to first think about the physical phenomena or instrumental sources responsible for the noise in the data, and to choose covariance functions that are appropriate for each source. At the end of the day, Gaussian processes are just like any other model: you get nothing more out than what you put in!

3.1.10 Useful References

Here is a list of references I have compiled over recent months, with help from others and which I hope you will in turn find useful.

- The following are milestone papers that have brought Gaussian processes to the field of exoplanets:

  – Gibson et al. (2011) and Czekala et al. (2014) introduced GPs to transmission spectroscopy for the study of planetary atmospheres;
  – Foreman-Mackey et al. (2015), Crossfield et al. (2015), Foreman-Mackey et al. (2014), Ambikasaran et al. (2014), Aigrain et al. (2015) and Barclay et al. (2015) harnessed the power of machine learning to detrend *Kepler* and *K2* lightcurves;
  – Baluev (2013), Haywood et al. (2014) and Grunblatt et al. (2015) applied GPs to RV studies;

3.1 Gaussian Processes

- Roberts et al. (2012) is not a paper specific to exoplanets but it provides a very clear introduction to GPs.

The use of GPs in our field is growing fast; this is only a small selection of papers and is in no way exhaustive.
- C. E. Rasmussen and C. K. I. Williams, *Gaussian Processes for Machine Learning*, the MIT Press, 2006, ISBN 026218253X., 2006 Massachusetts Institute of Technology (online: www.GaussianProcess.org/gpml). This is the classic reference in which you will find all the equations and statistical jargon.
- If you wish to develop your GP intuition, here is a fantastic lecture on the nature of Gaussian processes by Prof. David MacKay, from the Engineering Department at the University of Cambridge. The first Sections of this introduction to Gaussian processes are based on his lecture. I thoroughly recommend watching it! http://videolectures.net/gpip06_mackay_gpb/
- João Faria, from the Institute of Astrophysics and Space Sciences in Porto, has written a Fortran implementation of GPs, integrated in a platform for the analysis of RV data. The code is available at github.com/j-faria/OPEN.
- Daniel Foreman-Mackey at the department of Astronomy in New York University has written a lot of useful code in Python and C++, and has made it publicly available at: http://dan.iel.fm/research/.
- Finally, Dr. Suzanne Aigrain and her group have given many talks and posters about GPs, and some of their slides can be found online.

3.2 Monte Carlo Markov Chain (MCMC)

In this part of the chapter, I describe my RV model and outline the MCMC fitting procedure that I apply to determine the best fit parameters and their uncertainties.

3.2.1 Modelling Planets

The orbit of each planet is assumed Keplerian. I model them as follows:

$$\Delta RV_k(t_i) = K_k \big[\cos(\nu_k(t_i, t_{peri_k}, P_k) + \omega_k) + e_k \cos(\omega_k) \big]. \quad (3.18)$$

The period of the orbit of planet k is given by P_k, and its semi-amplitude is K_k. $\nu_k(t_i, t_{peri_k})$ is the true anomaly[4] of planet k at time t_i, and t_{peri_k} is the time of

---

[4]The true anomaly is "the angle between the direction of periastron and the current position of the planet measured from the barycentric focus of the ellipse" (Perryman 2011).

periastron. Because it is difficult to constrain the argument of periastron for planets in low-eccentricity orbits, we introduce two parameters C_k and S_k (Ford 2006). They are related to the eccentricity e_k of the planet's orbit and the argument of periastron ω_k as follows:

$$C_k = \sqrt{e_k} \cdot \cos(\omega_k), \qquad (3.19)$$

$$S_k = \sqrt{e_k} \cdot \sin(\omega_k). \qquad (3.20)$$

The use of the square root imposes a uniform prior on e_k, reducing the bias towards high eccentricities typically seen when defining C_k and S_k as $e_k \cos(\omega_k)$ and $e_k \sin(\omega_k)$ (see Sect. 3.2.4 for more detail on priors).

The eccentricity is defined as:

$$e_k = S_k^2 + C_k^2, \qquad (3.21)$$

and the argument of periastron is:

$$\omega_k = \tan^{-1}(S_k/C_k). \qquad (3.22)$$

3.2.2 Modelling Stellar Activity

My activity model is based on a Gaussian process trained on the off-transit variations in the star's lightcurve, with the quasi-periodic covariance function presented in Sects. 3.1.3 and 3.1.5.

In my analysis of the CoRoT-7 system, which was observed simultaneously with CoRoT and HARPS in 2012 (see Chap. 4), I was able to apply the *FF'* method of Aigrain et al. (2012) to model the suppression of convective blueshift and the flux blocked by starspots on the rotating stellar disc (see Chap. 2). I then used another Gaussian process with the same covariance properties to account for other activity-induced signals, such as photospheric inflows towards active regions or limb-brightened facular emission that is not spatially associated with starspots (Haywood et al. 2014).

For the vast majority of stars, however, we cannot obtain contemporaneous high precision photometric and spectroscopic observations—it is either too expensive or impractical in terms of telescope time, or space-based, high precision photometry is not available for the system considered. In these cases I use a Gaussian process on its own to account for all activity-induced RV signals (see my analyses of the Kepler-10 and Kepler-78 systems in Chap. 4).

3.2 Monte Carlo Markov Chain (MCMC)

3.2.2.1 Evaluating the FF' Activity Basis Functions

In order to calculate the FF' activity basis functions $\Delta RV_{\text{rot}}(t)$ and $\Delta RV_{\text{conv}}(t)$, the flux F at the time of each RV point has to be interpolated from the stellar lightcurve. I do this by training a Gaussian process on the lightcurve, which then allows me to predict the flux at each time of RV observation. I also interpolate the stellar flux at times $t + \Delta t$ and $t - \Delta t$, in order to compute the first time derivative of the flux:

$$F'(t) = \frac{F(t + \Delta t) - F(t - \Delta t)}{2\Delta t}. \tag{3.23}$$

This allows me to compute $\Delta RV_{\text{rot}}(t)$ and $\Delta RV_{\text{conv}}(t)$.

3.2.2.2 An Additional Activity Basis Function

I account for activity-related signals (not modelled by the FF' terms, if I am also applying this method) by introducing an activity basis function that takes the form of a GP. As I already discussed in Sect. 3.1.5, I implicitly assume that this GP has the same quasi-periodic covariance properties as the lightcurve. The basic concept is summarised in the diagram in Fig. 3.8. The amplitude of the GP, θ_1 is a free parameter in my total RV model (see Sect. 3.2.3). The other hyperparameters, θ_2, θ_3

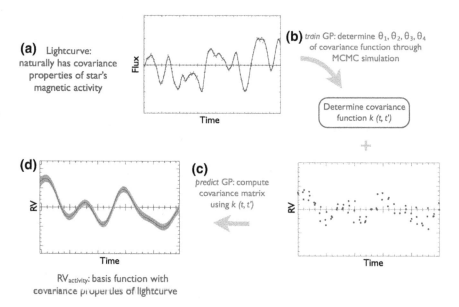

Fig. 3.8 Diagram outlining the 4-step procedure to follow in order to absorb potential activity-related red noise RV residuals using a Gaussian process that has the covariance properties of the *lightcurve*

and θ_4 are equal to the respective hyperparameters obtained when training a GP on the lightcurve. As the stellar activity mostly generates low-frequency signals, I refer to this activity term as $\Delta RV_{\text{rumble}}$.

3.2.2.3 Activity Model

The total RV perturbation $\Delta RV_{\text{activity}}$ induced by stellar activity is then:

$$\Delta RV_{\text{activity}} = A\Delta RV_{\text{rot}} + B\Delta RV_{\text{conv}} + \Delta RV_{\text{rumble}}, \qquad (3.24)$$

where A and B are scaling factors, and the amplitude of $\Delta RV_{\text{rumble}}$ is controlled by the hyperparameter θ_1 of Eq. 3.14.

3.2.3 Total RV Model

My final model consists of the three basis functions for the stellar activity as well as a Keplerian for each one of n_{pl} planets:

$$\Delta RV_{\text{tot}}(t_i) = RV_0 + \Delta RV_{\text{activity}}(t_i, A, B, \Psi_0, \theta_1)$$
$$+ \sum_{k=1}^{n_{pl}} K_k \big[\cos(\nu_k(t_i, t_{peri_k}, P_k) + \omega_k) + e_k \cos(\omega_k)\big], \qquad (3.25)$$

where RV_0 is a constant offset.

3.2.4 Choice of Priors

The priors I adopt for each parameter are given in Table 3.2. A Jeffreys prior for a parameter x has the form (Gregory 2007):

$$P(x|\mathcal{M}) = \frac{1}{x \ln(\frac{x_{\max}}{x_{\min}})}, \qquad (3.26)$$

where $x_{\min}$ and $x_{\max}$ are the lower and upper bounds of the parameter space that we choose to explore. For example, we apply this prior to parameters that represent timescales (eg. the planet orbital periods) because they follow a logarithmic scale; one year always seems longer to a child than to an adult because it represents a much larger fraction of their total life. In order to sample in an unbiased way, we must sample more sparsely at long timescales than at short timescales.

3.2 Monte Carlo Markov Chain (MCMC)

Table 3.2 Prior probability densities and ranges of the parameters modelled in the MCMC procedure

Symbol	Parameter	Prior
Systemic RV offset	RV_0	Uniform
Amplitude of GP	θ_1	Modified Jeffreys (σ_{RV})
Amplitude of $\Delta RV_{\rm rot}$	A	Modified Jeffreys (σ_{RV})
Amplitude of $\Delta RV_{\rm conv}$	B	Modified Jeffreys (σ_{RV})
Unspotted flux level	Ψ_0	Uniform [$\Psi_{\max}$, no upper limit]
Orbital period of non-transiting planet	$P_{\rm transiting}$	Gaussian ($P_{\rm transit}, \sigma_{P_{\rm transit}}$)
Transit ephemeris of non-transiting planet	$t_{0 \rm transiting}$	Gaussian ($t_{\rm transit}, \sigma_{t_{\rm transit}}$)
Orbital period of transiting planet	$P_{k \neq \rm transiting}$	Jeffreys
Transit ephemeris of transiting planet	$t_{0k \neq \rm transiting}$	Uniform
Planet RV amplitude	K_k	Modified Jeffreys (σ_{RV})
Planetary eccentricity (if transiting)	$e_{\rm transiting}$	Uniform $[0, 1 - \frac{R_\star}{a_h}]$
Planetary eccentricity (if not transiting)	$e_{k \neq \rm transiting}$	Uniform $[0, 1 - \frac{a_{k-1}}{a_k}(1 + e_{k-1})]$
Argument of periastron	ω_k	Uniform $[0, 2\pi]$

The knee of the modified Jeffreys prior is given in brackets. In the case of a Gaussian distribution, the terms within brackets represent the mean $\bar{x}$ and standard deviation σ. The terms within square brackets stand for the lower and upper limit of the specified distribution; if no interval is given, no limits are placed

A modified Jeffreys prior is given by:

$$P(x|\mathcal{M}) = \frac{1}{x + x_0} \frac{1}{x \ln(\frac{x_{\max}}{x_{\min}})}, \qquad (3.27)$$

where x_0 is the knee of the modified prior. This prior acts as a uniform prior when $K \ll \sigma_{RV}$, and as a Jeffreys prior for $K \gg \sigma_{RV}$. I choose the knee of the modified Jeffreys prior for the semi-amplitudes of the planets to be the mean estimated error of the RV observations, σ_{RV}. This ensures that the semi-amplitudes do not get overestimated in the case of a non-detection. I adopt the same modified Jeffreys prior for the amplitudes A and B of the FF' basis functions and the amplitude of the GP (θ_1). θ_1 is naturally constrained to remain low through the calculation of $\mathcal{L}$ (see the next Section). I constrain the orbital eccentricity of the innermost planet so that the planet's orbit remains above the stellar surface. I also impose a simple dynamical stability criterion on the outer planets by ensuring their eccentricities are such that the orbit of each planet does not cross that of its inner neighbour. I force the epochs of inferior conjunction of the outer non-transiting planets (corresponding to mid-transit for a 90° orbit) to occur close to the inverse variance-weighted mean date of the RV observations in order to ensure orthogonality with the orbital periods.

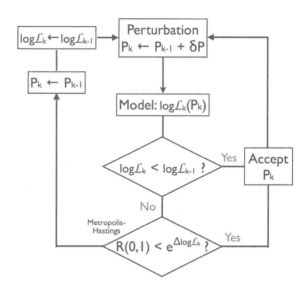

Fig. 3.9 The thought process of an MCMC simulation at each step

3.2.5 Fitting Procedure

The procedure is illustrated in the flow chart of Fig. 3.9. At every step of the chain, parameters A, B, Ψ_0, θ_1, RV_0, and the orbital elements of all planets are allowed to take a random jump in parameter space. The size of the step is equal to the size of the error bars of the parameters; at the start of the MCMC, these are mostly a guess, but the error bars (and therefore the step sizes) will be re-evaluated during the scaling phase, in order to ensure that the MCMC is taking steps of appropriate size. The hyperparameters θ_2, θ_3 and θ_4 are kept fixed as they are better constrained by the lightcurve than the RVs.

Likelihood The FF' activity basis functions (if used), together with the planet Keplerians and RV_0 are computed based on the present value of the parameters and subtracted from the data, in order to obtain the residuals r. The GP of the activity "rumble" term is then fitted to these residuals in order to absorb any signals with a frequency structure that matches that of the stellar activity. The likelihood $\mathcal{L}$ of the RV residuals is calculated at each step according to the following equation:

$$\ln \mathcal{L} = -\frac{n}{2}\ln(2\pi) - \frac{1}{2}\ln(\det \mathbf{K}) - \frac{1}{2}\underline{r}^T \cdot \mathbf{K}^{-1} \cdot \underline{r}, \tag{3.28}$$

which is very similar to Eq. 3.9.

Step acceptance or rejection The value of the likelihood is compared with that at the previous step: if the likelihood is higher, it means that this set of parameters provides a better fit than the previous set. The step is then accepted or rejected, the decision being made via the Metropolis-Hastings algorithm (Metropolis et al. 1953).

It allows some steps to be accepted when they yield a slightly poorer fit, in order to prevent the chain from becoming trapped in a local $\mathcal{L}$ maximum and instead explore the full parameter landscape. Ideally, the acceptance rate should be around 0.25; this ensures efficient and complete exploration of the parameter space (Ford 2006).

Burn-in phase and choice of parameter starting points If the planets in the system are not transiting and therefore do not know the orbital period and epoch, I usually make a generalised Lomb-Scargle periodogram (Zechmeister and Kürster 2009; Lomb 1976; Scargle 1982; see Sect. 2.3.2.1) to identify the strongest signals present in the dataset. The period, phase and approximate amplitude of these signals can then be used as starting points for the MCMC run. If the starting parameter values are wildly off from their true value, the MCMC simulation gets lost in obscure areas of parameter space and never converges (or takes a very long time to do so). An MCMC simulation should only be used in the aim of refining the optimal parameters of a model, and to estimate their error bars in a rigorous way; it is not intended for a first glimpse of what signals might be hiding in a dataset. In most cases, however, as long as some thought has been given to the choice of starting points, it will not affect the outcome of the chains. The initial burn-in phase is complete once $\mathcal{L}$ becomes smaller than the median of all previous $\mathcal{L}$ (Knutson et al. 2008).

Scaling phase After the burn-in phase, the MCMC goes through another set of steps, over which the standard deviations of all the parameters are then calculated. This phase allows the step sizes to be adjusted and should result in an optimum acceptance rate for the exploration phase.

Exploration phase and chain convergence The chain goes through a final set of steps in order to fully explore the parameter landscape in the vicinity of the maximum of $\mathcal{L}$. This last phase provides the joint posterior probability distribution of all parameters of the model. I check the good convergence of my code by applying the Gelman-Rubin criterion (Gelman et al. 2004; Ford 2006), which must be smaller than 1.1 to ensure that the chain has reached a stationary state. The best fit parameters are determined by taking the mean of the parameter chains over this phase, and their error bars can be obtained by calculating the standard deviations.

3.2.6 Care Instructions

It is important to look after your MCMC simulation to make sure it is in good health. Checking the acceptance rate is one way to assess whether the chains are behaving reasonably. If it is too low, the chains will move very slowly and will look like "slug trails"; if it is too high, they will not be able to close in on the likelihood maximum, and they will look like an excited bouncy bean. The parameter chains can be plotted as a function of step number. Another useful diagnostic is to plot them against each

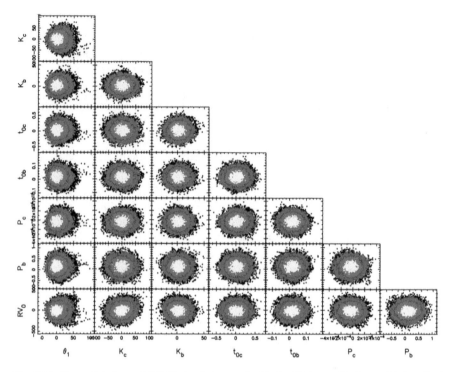

Fig. 3.10 Phase plots for an MCMC simulation on the Kepler-10 data, with the model described in Sect. 4.3 (next chapter). Points in *yellow*, *red* and *blue* are within the 1, 2 and 3-σ confidence regions, respectively. The scale of each axis corresponds to the departure of each parameter from its value at maximum likelihood. All are expressed in percent

other; such correlation plots for a healthy MCMC chain are shown in Fig. 3.10. The plots allow you to see whether any parameters are correlated with each other, which can considerably reduce the efficiency of an MCMC simulation. It is best practice to orthogonalise all your parameters before running the simulations.

3.3 Model Selection with Bayesian Inference

> In general, nature is more complicated than our model and known noise terms.
> P.C. Gregory (2007)

I run MCMC chains for several different models and select the best one according to the principles of Bayesian inference.

3.3 Model Selection with Bayesian Inference

3.3.1 Bayes' Factor

Given a dataset **y**, consider two models $\mathcal{M}_i$ and $\mathcal{M}_j$. In order to determine which one is the simplest but still gives the best fit to the data, we can compare the two models by estimating their posterior odds ratio:

$$\frac{P(\mathcal{M}_i|\mathbf{y})}{P(\mathcal{M}_j|\mathbf{y})} = \frac{Pr(\mathcal{M}_i)}{Pr(\mathcal{M}_j)} \cdot \frac{m(\mathbf{y}|\mathcal{M}_i)}{m(\mathbf{y}|\mathcal{M}_j)}, \quad (3.29)$$

where the first factor on the right side of the equation is the prior odds ratio. For the planetary systems I usually explore, all models that are tested have the same prior information, so this ratio is just 1. This leaves us with the second part of the right side of the equation. It is the ratio of the marginal likelihoods m of each model, and is known as Bayes' factor.

The marginal likelihood m of a dataset **y** given a model $\mathcal{M}_i$ with a set of parameters θ_i can be written as:

$$m(\mathbf{y}|\mathcal{M}_i) = \int f(\mathbf{y}|\mathcal{M}_i, \theta_i)\, \pi_i(\theta_i|\mathcal{M}_i)\, d\theta_i, \quad (3.30)$$

where $f(\mathbf{y}|\mathcal{M}_i, \theta_i)$ is the likelihood function $\mathcal{L}$. The term $\pi_i(\theta_i|\mathcal{M}_i)$ accounts for the prior distribution of the parameters and can be incorporated as a penalty to $\mathcal{L}$. According to Chib and Jeliazkov (2001), it is possible to write:

$$m(\mathbf{y}|\mathcal{M}_i) = \frac{f(\mathbf{y}|\mathcal{M}_i, \theta_i)\, \pi(\theta_i|\mathcal{M}_i)}{\pi(\theta_i|\mathbf{y}, \mathcal{M}_i)}. \quad (3.31)$$

The denominator $\pi(\theta_i|\mathbf{y}, \mathcal{M}_i)$ is the posterior ordinate, which we estimate using the posterior distributions of the parameters resulting from MCMC chains.

3.3.2 Posterior Ordinate

According to Chib and Jeliazkov (2001), the posterior ordinate $\hat{\pi}(\theta_i|\mathbf{y})$ can be evaluated by comparing the mean transition probability for a series of M jumps from any given θ_i to a reference θ_*, to the mean acceptance value for a series of J transitions *from* θ_*. This can be written as:

$$\hat{\pi}(\theta_*|\mathbf{y}) = \frac{M^{-1}\sum_{i=1}^{M}\alpha(\theta_i, \theta_*|\mathbf{y}) \cdot q(\theta_i, \theta_*|\mathbf{y})}{J^{-1}\sum_{j=1}^{J}\alpha(\theta_*, \theta_j|\mathbf{y})}, \quad (3.32)$$

where $\alpha(\theta_i, \theta_*|\mathbf{y})$ is the acceptance probability of the chain from one parameter set θ_i to another set θ_*. The proposal density $q(\theta_i, \theta_*|\mathbf{y})$ from one step θ_i to another θ_* is equal to:

$$q(\theta_i, \theta_*|\mathbf{y}) = \exp\left[-\sum_{k=1}^{K}\left(\frac{\theta_i - \theta_*}{\sigma_{\theta_i}}\right)^2/2\right]. \quad (3.33)$$

The summation inside the exponential term is carried out over all K parameters of the model, in other words over each parameter contained within a set θ.

If we choose θ_* to be the best parameter set of the whole MCMC chain, then the acceptance probability $\alpha(\theta_i, \theta_*|\mathbf{y})$ is 1, and Eq. 3.32 is much simplified.

3.3.3 Marginal Likelihood

I obtain $\mathcal{L}_{ML}$ by subtracting the posterior ordinate from the maximum likelihood value of the whole MCMC chain:

$$\log \mathcal{L}_{ML} = \log \mathcal{L}_{best} - \log \hat{\pi}. \quad (3.34)$$

When the number of model parameters becomes very large, the summation on the numerator of Eq. 3.32 is dominated by a relatively small fraction of points in the Markov chain that happen to lie close to the maximum likelihood value. A large number of trials is therefore needed to arrive at a reliable estimate of $\hat{\pi}$. I estimate the uncertainty in the posterior ordinate by running the chains several times and determining the variance empirically.

Once $\mathcal{L}_{ML}$ is known we can compute Bayes' factor for a pair of models. The posterior ordinate acts to penalise models that have too many parameters. Jeffreys (1961) found that the evidence in favour of a model is decisive if Bayes' factor exceeds 150, strong if it is in the range of 150–20, positive for 20–3 and not worth considering if lower than 3.

Concluding Note

Now that we have a recipe to detect planets around active stars, we can go look for them! In the next chapter, I present my analysis of the CoRoT-7, Kepler-78 and Kepler-10 systems.

References

Aigrain S, Hodgkin ST, Irwin MJ, Lewis JR, Roberts SJ (2015) Monthly notices of the royal astronomical society 447:2880
Aigrain S, Pont F, Zucker S (2012) Monthly notices of the royal astronomical society 419:3147

References

Ambikasaran S, Foreman-Mackey D, Greengard L, Hogg DW, O'Neil M (2014) Numerical analysis, pre-print (arXiv:1403.6015)
Baluev RV (2013) Monthly notices of the royal astronomical society 429:2052
Barclay T, Endl M, Huber D, Foreman-Mackey D, Cochran WD, MacQueen PJ, Rowe JF, Quintana EV (2015) Astrophys J 800:46
Chib S, Jeliazkov I (2001) American statistical association portal : marginal likelihood from the metropolis–hastings output. J Am Stat Assoc, 96:453
Crossfield IJM et al (2015) Astrophys J, pre-print (arXiv:1501.03798)
Czekala I, Andrews SM, Mandel KS, Hogg DW, Green GM (2014) Astrophys J, pre-print (arXiv:1412.5177)
Ford EB (2006) Astrophys J 642:505
Foreman-Mackey D, Hogg DW, Morton TD (2014) Astrophys J 795:64
Foreman-Mackey D, Montet BT, Hogg DW, Morton TD, Wang D, Schölkopf B (2015) Astrophys J, pre-print (arXiv:1502.04715)
Gelman A, Carlin JB, Stern HS, Rubin DB (2004) Bayesian data analysis. Chapman and Hall, London. (CRC, Boca Raton)
Gibson NP, Aigrain S, Roberts S, Evans TM, Osborne M, Pont F (2011) Monthly notices of the royal astronomical society 419:2683
Gregory PC (2007) Monthly notices of the royal astronomical society 381:1607
Grunblatt SK, Howard AW, Haywood RD (2015) Astrophys J, pre-print (arXiv:1501.00369)
Haywood RD et al (2014) Monthly notices of the royal astronomical society, 443(3):2517–2531
Jeffreys SH (1961) The theory of probability. Oxford University Press, Oxford
Knutson HA, Charbonneau D, Allen LE, Burrows A, Megeath ST (2008) Astrophys J 673:526
Lomb NR (1976) Astrophys Space Sci 39:447
Metropolis N, Rosenbluth AW, Rosenbluth MN, Teller AH, Teller E (1953) J Chem Phys 21:1087
Perryman M (2011) The exoplanet handbook by michael perryman
Press WH, Flannery BP, Teukolsky SA (1986) Numerical recipes. The art of scientific computing, Vol 1. Cambridge University Press, Cambridge
Rasmussen CE, Williams CKI (2006) Gaussian processes for machine learning. MIT Press, Massachusetts
Roberts S, Osborne M, Ebden M, Reece S, Gibson N, Aigrain S (2012) Philosophical transactions of the royal society a: mathematical. Phys Eng Sci 371:20110550
Scargle JD (1982) Astrophys J 263:835
Zechmeister M, Kürster M (2009) Astron Astrophys 496:577

Chapter 4
Application to Observations of Planet-Hosting Stars

In this chapter, I present the work I have done towards characterising three planetary systems: CoRoT-7, Kepler-78 and Kepler-10. CoRoT-7 is an active star host to a small hot Neptune and the first Earth-size exoplanet ever discovered. Kepler-78 is an active star orbited by an extremely close-in hot, Earth-mass planet. Kepler-10 is old and quiet, and harbours two transiting planets—one a super-Earth, the other a rocky world the size of Neptune whose discovery challenges our theories of planet formation. All three of these systems were first discovered via the transit method, by the CoRoT and *Kepler* space missions. They were followed up with HARPS, HARPS-N and HIRES in order to obtain a precise mass determination of the planets present in these systems.

This chapter uses material from, and is based on, Haywood et al. (2014), MNRAS, 443, 2517 and my own contributions to Dumusque et al. (2014), ApJ, 789, 154, and Grunblatt et al. (2015), ApJ, 808, 127.

CoRoT-7b: the first transiting super-Earth with a measured radius! Author: Fabien Catalano

4.1 CoRoT-7

Since the discovery of the super-Earth CoRoT-7b, several investigations have yielded different results for the number and masses of planets present in the system, mainly owing to the star's high level of activity. This system has a long history, which I present in the next section, before I report on my own analysis in the following sections.

4.1.1 History of the System

In July 2009, Léger et al. (2009) announced the discovery of a transiting planet CoRoT-7b, the first Super-Earth with a measured radius found by the CoRoT space mission. At the time, it had the smallest exoplanetary radius ever measured, $R_b = 1.68 \pm 0.09 R_\oplus$. CoRoT-7 is relatively bright (V = 11.7) but has fairly high activity levels, meaning that for a long time the number of planets detected around it and their precise physical parameters remained in debate.

Following this discovery, a 4-month intensive HARPS campaign was launched in order to measure the mass of CoRoT-7b. The results of this run are reported in

4.1 CoRoT-7

Queloz et al. (2009). They expected the RV variations to be heavily affected by stellar activity, given the large modulations in the CoRoT photometry. The star's lightcurve (2008–2009 CoRoT run) shows modulations due to starspots of up to 2%, which tells us that CoRoT-7 is more active than the Sun, whose greatest recorded variations in irradiance are of 0.34% (Kopp and Lean 2011). Indeed, a few simultaneous photometric measurements from the Euler Swiss telescope confirmed that CoRoT-7 was very spotted throughout the HARPS run. In order to remove the activity-induced RV variations from the data, Queloz et al. (2009) applied a pre-whitening procedure followed by a harmonic decomposition (see Sect. 2.2.3). For the prewhitening, the period of the stellar rotation signal was identified by means of a Fourier analysis, and a sine fit with this period was subtracted from the data. This operation was applied to the residuals to remove the next strongest signal, and so on until the noise level was reached. All the signals detected with this method were determined to be associated with harmonics of the stellar rotation period, except two signals at 0.85 and 3.69 days. The RV signal at 0.85 days was found to be consistent with the CoRoT transit ephemeris, thus confirming the planetary nature of CoRoT-7b. Its mass was determined to be 4.8 ± 0.8 $M_\oplus$. In order to assess the nature of the signal at 3.69 days, Queloz et al. (2009) used a harmonic decomposition to create a high pass filter: the RV data were fitted with a Fourier series comprising the first three harmonics of the stellar rotation period, within a time window sliding along the data. The length of this window (coherence time) was chosen to be 20 days, so that any signals varying over a longer timescale were filtered out—starspots typically have lifetimes of about a month (Schrijver 2002; Hussain 2002, see Chap. 2). The harmonically filtered data were found to contain a strong periodic signal at 3.69 days, which was attributed to the orbit of CoRoT-7c, another super-Earth with a mass of 8.4 ± 0.9 $M_\oplus$.

A few months later, Bruntt et al. (2010) re-measured the stellar radius with improved stellar analysis techniques, which led to a slightly smaller planetary radius for CoRoT-7b than initially found, of $1.58 \pm 0.10 R_\oplus$.

A separate investigation was later carried out by Lanza et al. (2010). The stellar induced RV variations were synthesized based on a fit to the CoRoT lightcurve, which was computed according to a maximum entropy spot model (Lanza et al. 2009, 2011). The existence of the two planets was then confirmed by demonstrating that the activity-induced RV variations did not contain any spurious signals at the orbital periods of the two planets, with an estimated false alarm probability of less than 10^{-4}.

In another analysis, Hatzes et al. (2010) applied a prewhitening procedure to the full width at half-maximum (FWHM), bisector span and Ca II H&K line emission derived from the HARPS spectra and cross-correlation analyses. These quantities vary according to activity only, and are independent of planetary orbital motions (see Sect. 2.2.1.1). No significant signals were found in any of these indicators at the periods of 0.85 and 3.69 days. Furthermore, they investigated the nature of a signal found in the RV data at 9.02 days. It had been previously detected by Queloz et al. (2009) but had been attributed to a "two frequency beating mode" resulting from an amplitude modulation of a signal at a period of 61 days. This is close to twice the stellar rotation period so it was deemed to be activity related. Hatzes et al. found no trace of a signal at 9.02 days in any of the activity indicators. They thus suggested this RV signal could be attributed to a third planetary companion with a mass of 16.7 ± 0.42 $M_\oplus$. They also confirmed the presence of CoRoT-7b and CoRoT-7c, but found different masses than calculated by Queloz et al. (2009). This was inevitable since the derived masses of planets are intimately connected with the methods used to mitigate the effects of stellar activity on the RV data.

Hatzes et al. (2010, 2011) developed a very simple method to remove stellar activity-induced RV variations, to obtain a more accurate mass for CoRoT-7b. The method relied on making several well-separated observations on each night, which was the case for about half of the HARPS data. Under the assumption that the variations due to activity and other planets were negligible during the span of the observations on each night, it was possible to fit a Keplerian orbit assuming that the velocity zero-point differs from night to night but remained constant within each night (see Sect. 2.2.2). Hatzes et al. (2010) report a mass of CoRoT-7b of 6.9 ± 1.4 $M_\oplus$ and the second analysis (Hatzes et al. 2011) yields a mass of 7.42 ± 1.21 $M_\oplus$, which is consistent.

Pont et al. (2010) carried out an analysis based on a maximum entropy spot model (similar to Lanza et al. 2010) which made use of many small spots as opposed to few large spots. The model was constrained using FWHM and bisector information. A careful examination of the residuals of the activity and planet models led to the authors to add an additional noise term in order to account for possible systematics beyond the formal RV uncertainties. Pont et al. (2010) argued that CoRoT-7b was detected in the RV data with much less confidence than in previous analyses, and reported a mass of 2.3 ± 1.8 $M_\oplus$ detected at a 1.2σ level. Furthermore, they argued that the RV data were not numerous enough and lacked the quality required to look for convincing evidence of additional companions.

Boisse et al. (2011) applied their SOAP tool Boisse et al. (2011) to the CoRoT-7 system. This program simulates spots on the surface of a rotating star and then uses this model to compute the activity-induced RV variations of the star. With this technique, they obtained mass estimates for CoRoT-7b and CoRoT-7c. They judge that their errors are underestimated and suggest adding a noise term of 1.5 m $\cdot$ s^{-1} to account for activity-driven RV variations. Their mass estimate for CoRoT-7b was in agreement with the value reported by Queloz et al. (2009) but they found a slightly higher value for the mass of CoRoT-7c.

Ferraz-Mello et al. (2011) constructed their own version of the high-pass filter employed by Queloz et al. (2009) in order to test the validity of this method and estimated masses for CoRoT-7b and 7c. They compared it to the method used by Hatzes et al. (2010, 2011) and to a pure Fourier analysis. They concluded the method was robust, and obtained revised masses of 8.0 ± 1.2 $M_\oplus$ for CoRoT-7b and 13.6 ± 1.4 $M_\oplus$ for CoRoT-7c, but made no mention of CoRoT-7d.

The analysis by Lanza et al. (2010), which makes use of the CoRoT lightcurve (Léger et al. 2009) to model the activity-induced RV variations, and those by Pont et al. (2010) and Boisse et al. (2011), which rely on the tight correlation between the FWHM and the simultaneous Euler photometry (Queloz et al. 2009), could be much improved with simultaneous photometric and RV data (see Lanza et al., in prep.).

4.1 CoRoT-7

The spot activity on CoRoT-7 changes very rapidly and it is therefore not possible to deduce the form of the activity-driven RV variations from photometry taken up to a year before the RV data.

In the next section, I introduce the new simultaneous photometric and RV observations obtained in 2012 January with the CoRoT satellite and HARPS spectrograph. I implement my model in Sect. 4.1.4, and discuss the outcomes in Sect. 4.1.5.

4.1.2 Observations

4.1.2.1 HARPS Spectroscopy

Radial velocities The CoRoT-7 system was observed with the HARPS instrument on the ESO 3.6 m telescope at La Silla, Chile for 26 consecutive clear nights from 2012 January 12 to February 6, with multiple well-separated measurements on each night. The 2012 RV data, shown in Fig. 4.1, were reprocessed in the same way as the 2008–2009 data (Queloz et al. 2009) using the HARPS data analysis pipeline. The cross-correlation was performed using a K5 spectral mask. The data are available in Table A1 of Haywood et al. (2014). The median, minimum and maximum signal-to-noise ratio of the HARPS spectra at central wavelength 556.50 nm are 44.8, 33.8 and 56.2, respectively. The RV variations during the second run, shown in Fig. 4.1 have a smaller amplitude than during the first HARPS campaign, implying that the star has become less active than it was in 2008–2009.

Time Series of Trailed Spectra I grouped the reprocessed cross-correlation functions (CCFs) into nightly averages, and subtracted the mean CCF (calculated over the full run) in order to obtain the residual perturbations to each line profile. I then stacked each of these residuals on top of one another as a function of time. The resultant trail of spectra is shown in Fig. 4.2. The bright trails are produced by starspots or groups of faculae drifting across the surface as the star rotates.

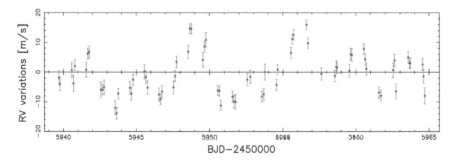

Fig. 4.1 RV observations of CoRoT-7, made in January 2012 with HARPS

Fig. 4.2 Trailed CCFs: the average line profile is subtracted from the individual profiles, which are then stacked vertically as a function of time. The two vertical dark lines represent gaps between groups of magnetically active regions crossing the stellar disc as the star rotates. The small variations along the horizontal scale within individual line profiles arise from the uniqueness of each pixel on the CCD

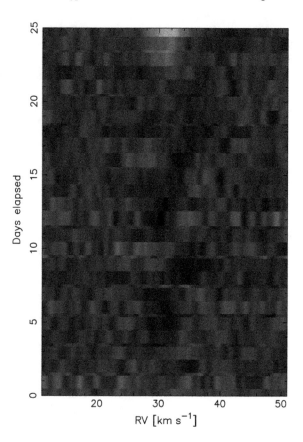

4.1.2.2 CoRoT Photometry

CoRoT-7 was observed with the CoRoT satellite (Auvergne et al. 2009) from 2012 January 10 to March 29. Figure 4.3 shows the part of the lightcurve which overlaps with the 2012 HARPS run. Measurements were taken in CoRoT's high cadence mode (every 32 s). The data were reduced with the CoRoT imagette pipeline with an optimised photometric mask in order to maximise the signal-to-noise ratio of the lightcurve. Further details on the data reduction are given by Barros et al. (2014), who present a combined analysis of both CoRoT datasets. They derive the revised orbital period and epoch of first transit shown in Table 4.1. These values will be used as prior information in my MCMC simulations (see Sect. 4.1.4). I binned the data in blocks of 0.07 day, which corresponds to 6176 s and is close to the orbital period of the satellite of 6184 s (Auvergne et al. 2009) in order to average the effects of all sources of systematic errors related to the orbital motion of CoRoT.

4.1 CoRoT-7

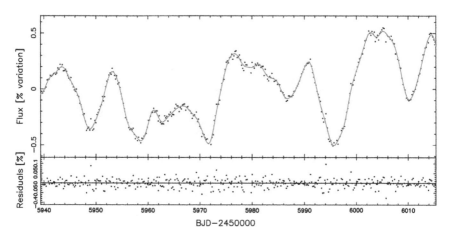

Fig. 4.3 *Upper panel* CoRoT-7 lightcurve over the span of the 2012 RV run, with my photometric fit at each RV observation overplotted as the *blue curve*. *Lower panel* Residuals of the fit

Table 4.1 Transit and star information based on both CoRoT runs (results from Barros et al. 2014)

CoRoT-7	
Spectral type	G9V
Mass	$0.913 \pm 0.017\,M_\odot$
Radius	$0.820 \pm 0.019\,R_\odot$
Age	$1.32 \pm 0.76\,\mathrm{Gyr}$
CoRoT-7b	
Orbital period	$0.85359165 \pm 5.6 \times 10^{-7}$ day
Transit ephemeris	$2454398.07694 \pm 6.7 \times 10^{-4}\,\mathrm{HJD}$
Transit duration	$1.42 \pm 0.15\,\mathrm{h}$
Orbital inclination	$80.78^{+0.51}_{-0.23}$ deg
Radius	$1.585 \pm 0.064\,R_\oplus$

4.1.3 Preliminary Periodogram Analysis

In order to determine an appropriate set of parameters as a starting point for my MCMC analysis, I made a periodogram of the 2012 RV data, shown in Fig. 4.4 (see Sect. 2.3.2.1). The stellar rotation period and its harmonics are marked by the red lines (solid and dashed, respectively). Because the orbital period of CoRoT-7b is close to 1 day, its peak in the periodogram is hidden amongst the aliases produced by the two strong peaks at 3.69 and 8.58 days. The peak at 3.69 days matches the period for CoRoT-7c of Queloz et al. (2009). We see another strong peak at a period of 8.58 days, which is close to the period found by Lanza (in prep.) of 8.29 days for the candidate planet signal CoRoT-7d, and about half a day shorter than that determined

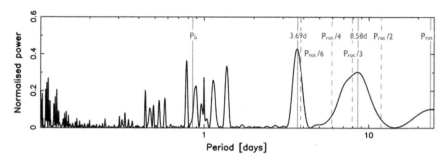

Fig. 4.4 Generalised Lomb–Scargle periodogram of the 2012 RV dataset. The stellar rotation fundamental, $P_{\rm rot}$, and harmonics are represented with *solid* and *dashed lines*, respectively. Also shown are the orbital period of CoRoT-7b derived from the transit analysis of Barros et al. (2014), $P_{\rm b}$, and the periods of the two strong peaks at 3.69 and 8.58 days

by Hatzes (in prep.) based on the same dataset. The periodogram shows that this peak is very broad and spans the whole 8–9 days range. Several stellar rotation harmonics are also present within this range, so at this stage I cannot conclude on the nature of this signal (this is discussed further in Sect. 4.1.5.3).

4.1.4 MCMC Analysis

4.1.4.1 RV Model

The planet orbits are modelled as Keplerians, while the activity model is based on a Gaussian process with a quasi-periodic covariance function trained on the off-transit variations in the star's lightcurve (see next section). I then use this Gaussian process in two ways:

(a) I model the suppression of convective blueshift and the flux blocked by starspots on a rotating star, via the FF' method of Aigrain et al. (2012). This method is explained in detail in Sect. 2.2.5 of Chap. 2.
(b) I use another Gaussian process with the same covariance properties to account for other activity-induced signals, such as photospheric inflows towards active regions or limb-brightened facular emission that is not spatially associated with starspots (Haywood et al. (2014), see Sect. 2.1.5.3).

The total RV model has the form:

$$\Delta RV_{\rm tot}(t_i) = RV_0 + A\Delta RV_{\rm rot} + B\Delta RV_{\rm conv} + \Delta RV_{\rm rumble}$$
$$+ \sum_{k=1}^{n_{pl}} K_k \big[\cos(\nu_k(t_i, t_{peri_k}, P_k) + \omega_k) + e_k \cos(\omega_k)\big], \quad (4.1)$$

4.1 CoRoT-7

where all the symbols have their usual meaning (refer to Sect. 3.2.3). The stellar radius $R_\star$, which is needed to calculate the $\Delta RV_{\rm rot}$ basis function of the FF' method, is set to the value determined by Barros et al. (2014), given in Table 4.1. The second FF' basis function, $\Delta RV_{\rm conv}$, depends on the difference between the convective blueshift in the unspotted photosphere and that within the magnetised area (δV_c) and the ratio of this area to the spot surface (κ). We do not know their values in the case of CoRoT-7 so they will be absorbed into the scaling constant B.

4.1.4.2 Gaussian Process

I interpolated the flux from the CoRoT lightcurve at the time of each RV point using a Gaussian process with a quasi-periodic covariance function:

$$k(t, t') = \eta_1^2 \cdot \exp\left(-\frac{(t-t')^2}{2\eta_2^2} - \frac{2\sin^2(\frac{\pi(t-t')}{\eta_3})}{\eta_4^2}\right). \tag{4.2}$$

It is the same as Eq. 3.14 that I introduced in the previous chapter. The shape of this covariance function reflects the quasi-periodic nature of the CoRoT lightcurve, as evolving active regions come in and out of view.

The hyperparameters are determined via the MCMC simulation described in Sect. 3.1.6.

1. Amplitude of the Gaussian process, η_1;
2. Timescale for growth and decay of active regions, η_2: I found it to be $\eta_2 = 20.6 \pm 2.5$ days. This implies that the active regions on the stellar surface evolve on timescales similar to the stellar rotation period;
3. Stellar rotation period, η_3: I computed the discrete autocorrelation function (Edelson and Krolik 1988) of the lightcurve (it is displayed in the second panel of Fig. 3.7 in the previous chapter). I find $P_{\rm rot} = 23.81 \pm 0.03$ days, which is consistent with the estimate of Léger et al. (2009) of about 23 days. I applied this value as a Gaussian prior in the MCMC simulation I ran to determine the other hyperparameters;
4. Finally, η_4 determines how smooth the fit is.

The fit is shown in the top panel of Fig. 4.3. The residuals of the fit shown in the bottom panel show no correlated noise and have an RMS scatter of 0.02 %. The parameters of the RV model are then fitted via the MCMC procedure that I presented in Sect. 3.2 of the previous chapter.

4.1.5 Results and Discussion

4.1.5.1 Justification for the Use of a Gaussian Process in Addition to the FF' method

Initially, I used the FF' basis functions on their own to account for activity-induced signals in the RVs. However, it quickly became apparent that an additional term is needed to account for all slowly-varying signals. I find that an RV model including a Gaussian process with a quasi-periodic covariance structure is the only model that yields uncorrelated, flat residuals. Regardless of the number of planets modelled, without the inclusion of this Gaussian process the residuals always display correlated behaviour. Figure 4.5 shows the residuals remaining after fitting the orbits of CoRoT-7b, CoRoT-7c and a third Keplerian, and the two basis functions of the FF' model. We see that even the addition of a third Keplerian does not absorb these variations, which appear to be quasi-periodic. Also, I note that a Gaussian process with a less complex, square exponential covariance function does not fully account for correlated residuals in either a 2- or 3-planet model. A comparison between a model with 2 planet orbits, the FF' basis functions and a Gaussian process that has square exponential or quasi-periodic covariance properties yields a Bayes factor of 3×10^6 in favour of the latter. This implies that the active regions on the stellar surface do remain, in part, from one rotation to the next.

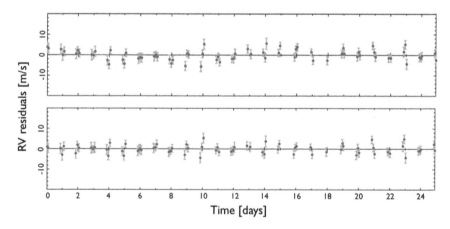

Fig. 4.5 *Top* RV residuals remaining after fitting a 3-planet $+ FF'$ activity functions model. They contain quasi-periodic variations, and show the need to use a red noise "absorber" such as a Gaussian process. *Bottom* RV residuals after including a Gaussian process with a quasi-periodic covariance function in our RV model. The RMS of the residuals, now uncorrelated, is $1.96 \text{ m} \cdot \text{s}^{-1}$ which is at the level of the error bars of the data

4.1.5.2 Identifying the Best Model Using Bayesian Model Selection

I ran MCMC simulations for models with 0 (activity only), 1, and 2 planets. I estimated the marginal likelihood of each model from the MCMC samples using the method of Chib and Jeliazkov (2001), which I described in Sect. 3.3. The marginal likelihoods are listed in the second to last row of Table 4.2. I also tested a 3-planet model, which I discuss in the next section.

The 2-planet model is preferred over the activity-only and 1-planet model (see the first three columns in Table 4.2). I also found that a 2-planet model with free orbital eccentricities is preferred over a model with forced circular orbits by a Bayes' factor of $5 \cdot 10^3$ (see Sect. 3.3). The model with forced circular orbits is penalised mostly because of the non-zero eccentricity of CoRoT-7c. Indeed, keeping e_b fixed to zero while letting e_c free yields a Bayes' factor of 270 (over a model with both orbits circular), while the Bayes' factor between models with e_b fixed or free (e_c free in both cases) is only 36. A model with no planets, consisting solely of the FF' basis functions and a quasi-periodic Gaussian process (Model 0) is severely penalised; this attests that models with the covariance properties of the stellar activity do not absorb the signals of planets b and c.

4.1.5.3 CoRoT-7d or Stellar Activity?

I investigated the outputs of 3-planet models in order to look for the 9-day signal present in the 2009 RV data (Queloz et al. 2009; Hatzes et al. 2010), whose origin has been strongly debated (cf. Sect. 4.1.1 and references therein).

First, I fitted a model comprising three Keplerians, the FF' basis functions and an additional Gaussian process with a quasi-periodic covariance function. I recover the orbits of the two inner planets but do not detect another signal with any significance. The residuals are uncorrelated and at the level of the error bars. I then constrained the orbital period of the third planet with a Gaussian prior centred around the period recently reported by Tuomi et al. (2014) at $P_d = 8.8999 \pm 0.0082$ days, and imposed a Gaussian prior centred at 2455949.97 ± 0.44 BJD on the predicted time of transit (which corresponds to the phase I determined based on the orbital period of Tuomi et al. (2014). I recover a signal which corresponds to a planet mass of 13 ± 5 M$_\oplus$ and is in agreement with the mass proposed by Tuomi et al. (2014). However, the log marginal likelihood of this model is -192.5 ± 0.7; this is lower than the log marginal likelihood of the 2-planet model (Model 2, $\log \mathcal{L}_{ML} = -190.1 \pm 0.7$), which suggests that the addition of an extra Keplerian orbit at 9 days is not justified in view of the improvement to the fit.

Since this orbital period is very close to the second harmonic of the stellar rotation, it is plausible that the Gaussian process could be absorbing some or all of the signal produced by a planet's orbit at this period. In order to test whether this is the case, I took the residuals of Model 2 and injected a synthetic sinusoid with the orbital parameters of planet d reported by Tuomi et al. (2014). I fitted this fake dataset with a model consisting of a Gaussian process (with the same quasi-periodic covariance

Table 4.2 Outcome of a selection of models: Model 0: stellar activity only, modelled by the FF' basis functions and a Gaussian process with a quasi-periodic covariance function; Model 1: activity and 1 planet; Model 2: activity and 2 planets; Model 2b: activity and 2 planets with eccentricities fixed to 0

	Model 0	Model 1	Model 2	Model 2b
Stellar activity				
A (m · s^{-1})	-0.36 ± 0.20	-0.35 ± 0.21	0.06 ± 0.13	0.06 ± 0.12
B (m · s^{-1})	0.84 ± 1.07	-0.35 ± 1.30	0.64 ± 0.28	0.49 ± 0.35
Ψ_0/Ψ_{max}	1.014 ± 0.013	1.014 ± 0.012	1.014 ± 0.012	1.014 ± 0.013
θ_1 (m · s^{-1})	75 ± 19	86 ± 20	7 ± 2	8 ± 2
CoRoT-7b				
P (days)		$0.85359165(6)$	$0.85359165(5)$	$0.85359163(6)$
t_0 (BJD—2450000)		$4398.0769(7)$	$4398.0769(8)$	$4398.0769(8)$
t_{peri} (BJD—2450000)		$4398.10(7)$	$4398.21(9)$	$4398.863(1)$
K (m · s^{-1})		3.95 ± 0.71	3.42 ± 0.66	3.10 ± 0.68
e		0.17 ± 0.09	0.12 ± 0.07	0 (fixed)
ω (°)		105 ± 61	160 ± 140	0 (fixed)
m ($M_\oplus$)		5.37 ± 1.02	4.73 ± 0.95	4.45 ± 0.98
ρ (g · cm^{-3})		7.51 ± 1.43	6.61 ± 1.33	6.21 ± 1.37
a (AU)		$0.017(1)$	$0.017(1)$	$0.017(1)$
CoRoT-7c				
P (days)			3.70 ± 0.02	3.68 ± 0.02
t_0 (BJD—2450000)			$5953.54(7)$	$5953.59(5)$
t_{peri} (BJD—2450000)			$5953.3(3)$	$5952.67(6)$
K (m · s^{-1})			6.01 ± 0.47	5.95 ± 0.48
e			0.12 ± 0.06	0 (fixed)
m ($M_\oplus$)			13.56 ± 1.08	13.65 ± 1.10
a (AU)			$0.045(1)$	$0.045(2)$
n_{obs}	71	71	71	71
n_{params}	5	10	15	11
$\log \mathcal{L}_{max}$	-237.6 ± 0.3	-223.6 ± 0.5	-188.0 ± 0.2	-196.28 ± 0.04
$\hat{\pi}$	0 ± 1	2 ± 1	2 ± 1	2.2 ± 0.8
$\log \mathcal{L}_{ML}$	-237 ± 1	-225 ± 1	-190.1 ± 0.7	-198.5 ± 0.8
Bayes' factor: $B_{k,2}$	4×10^{-21}	6×10^{-16}	–	2×10^{-4}
BIC	496.5	489.8	439.9	439.4

The numbers in brackets represent the uncertainty in the last digit of the value. Also given are the number of observations used (n_{obs}), the number of parameters in each model (n_{params}), the maximum likelihood ($\log \mathcal{L}_{max}$), the posterior ordinate ($\hat{\pi}$), the marginal likelihood ($\log \mathcal{L}_{ML}$) and the Bayesian Information Criterion (BIC) for each model. In the last row, each model is compared to Model 2 using Bayes' factor

4.1 CoRoT-7

function as before), a Keplerian and a constant offset. I find that the planet signal is completely absorbed by the Keplerian model, within uncertainties—the amplitude injected was 5.16 ± 1.84 m·s^{-1}, while that recovered is 4.97 ± 0.35 m·s^{-1}. This experiment attests that the likelihood of the model (see Eq. 3.28) acts to keep the amplitude of the Gaussian process as small as possible, in order to compensate for its high degree of flexibility, and allow other parts of the model to fit the data if they are less complex than the Gaussian process. I therefore conclude that if there were a completely coherent signal close to 9 days, it would be left out by the Gaussian process and be absorbed by the third Keplerian of the 3-planet model.

This signal therefore cannot be fully coherent over the span of the observations. Indeed, we see in the periodogram of the RV data in Fig. 4.4 that the peak at this period is broad. I note that despite the lower activity levels of the star in the 2012 dataset, the 9-day period is less well determined in this dataset than in the 2008–2009 one. This peak is also broader than we would expect for a fully coherent signal at a period close to 9 days with the observational sampling of the 2012 dataset. This is likely to be caused by variations in the phase and amplitude of the signal over the span of the 2012 data.

Based on the 2012 RV dataset, I do not have enough evidence to confirm the presence of CoRoT-7d as its orbital period of 9 days is very close to the second harmonic of the stellar rotation. Furthermore, the period measured for the 2009 dataset by Hatzes et al. (2010) $P_d = 9.021 \pm 0.019$ days is not precise enough to allow me to determine whether the signals from the two seasons are in phase, as was done in the case of α Centauri Bb by Dumusque et al. (2012). The cycle count of orbits elapsed between the two datasets is: $n = 1160/9.021 = 128.6$ orbits. The uncertainty is $n\,\sigma_{P_d}/P_d = n\,(0.019/9.021) = 0.27$ orbits. Although this 1-sigma uncertainty is less than one orbit, it is big enough to make it impossible to test whether the signal is still coherent. The most likely explanation, given the existing data, is that the 8–9 day signal seen in the periodogram of Fig. 4.4 is a harmonic of the stellar rotation.

4.1.5.4 Best RV Model: 2 Planets and Stellar Activity

Figure 4.6 shows each component of the total RV model plotted over the duration of the RV campaign. We see that the suppression of convective blueshift by active regions surrounding starspots has a much greater impact on RV than flux blocked by starspots; I discuss this further in Sect. 4.1.5.7.

Figure 4.7 shows Lomb–Scargle periodograms of the CoRoT 2012 lightcurve and the HARPS 2012 RV data. Panel (a) shows the periodogram of the full CoRoT 2012 lightcurve, while panel (b) represents the periodogram of the Gaussian process fit to the lightcurve sampled at the times of the HARPS 2012 RV observations. Both periodograms reveal a stronger peak at $P_{rot}/2$ than at P_{rot}, which indicates the presence of two major active regions on opposite hemispheres of the star. This is in agreement with the variations in the lightcurve in Fig. 4.3. Given that suppression of convective blueshift appears to be the dominant signal, we would expect a similar frequency structure to be present in the periodogram of the RV curve (panel (c)).

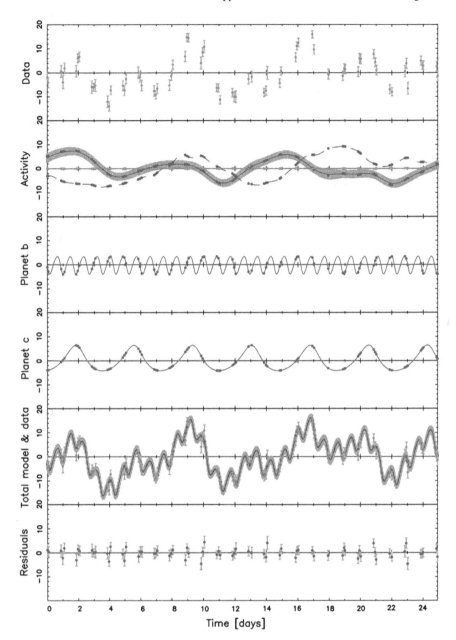

Fig. 4.6 Time series of the various parts of the total RV model for Model 2, after subtracting the star's systemic velocity RV_0. All RVs are in m · s^{-1}. Panel (**b**) $A\Delta RV_{\text{rot}}$ (*orange full line*), $B\Delta RV_{\text{conv}}$ (*purple dashed line*) and $\Delta RV_{\text{rumble}}$ (*blue full line* with *grey* error band). Panel (**e**) the total model (*red*), which is the sum of activity and planet RVs, is overlaid on top of the data (*blue points*). Subtracting the model from the data yields the residuals plotted in panel (**f**)

4.1 CoRoT-7

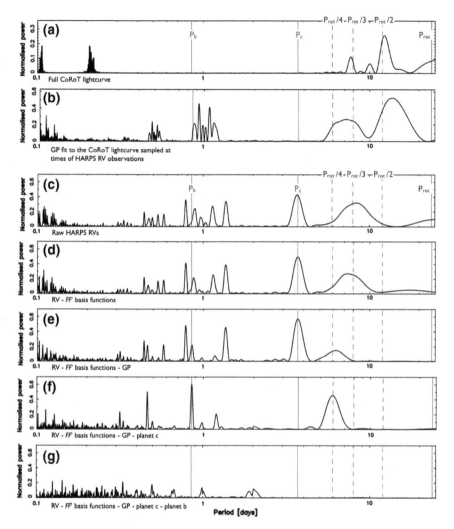

Fig. 4.7 Lomb–Scargle periodograms of: **a** the full 2012 CoRoT lightcurve; **b** the Gaussian process fit to the 2012 CoRoT lightcurve sampled at the times of RV observations; **c** the raw 2012 HARPS RV observations; **d** the RV data, from which I subtracted the FF' basis functions; **e** same as (**d**), with the Gaussian process also removed; **f** same as (**e**), with the signal of planet c removed; **g** same as (**f**), with planet b removed

Indeed, we see that the stellar rotation harmonics bracket the 6–10 day peak in the periodogram, which has significantly greater power than the fundamental 23-day rotation signal. In panel (d), I remove the two FF' basis functions. I then subtract the Gaussian process (panel (e)). We see that the Gaussian process absorbs most of the power present in the 6–10 day range. In panel (f), I have also subtracted the orbit of planet c. This removes the peaks at P_c and its 1-day alias at $\sim$1.37 days. The peak due

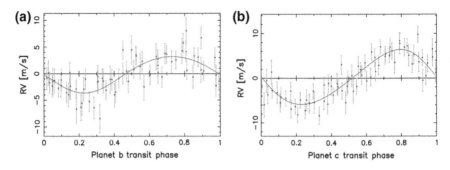

Fig. 4.8 *Panel* **a** Phase plot of the orbit of CoRoT-7b for Model 2, with the contribution of the activity and CoRoT-7c subtracted. *Panel* **b** Phase plot of the orbit of CoRoT-7c for Model 2, with the contribution of the activity and CoRoT-7b subtracted

to CoRoT-7b now stands out along with its 1-day alias at $P = 1/(1 - 1/P_b) \sim 5.82$ days and harmonics $P_b/2$ and $P_b/3$. Finally, I subtract the orbit of planet b, which leaves us with the periodogram of the residuals. We see that no strong signals remain except at the 1- and 2-day aliases arising from the window function of the ground-based HARPS observations.

There are no strong correlations between any of the parameters. The K amplitudes of planets b and c are found to be unaffected by the number of planets, choice of eccentric or circular orbits, or choice of activity model (all, some or none of $\Delta RV_{\text{activity}}$), even when I leave P_c unconstrained. The residuals, with an RMS scatter of 1.96 m · s^{-1}, are at the level of the error bars of the data and show no correlated behaviour, as seen in the bottom panel of Fig. 4.5.

4.1.5.5 CoRoT-7b

The orbital parameters of CoRoT-7b are listed in the third column of Table 4.2. The orbital eccentricity of 0.12 ± 0.07 is detected with a low significance and is compatible with the transit parameters determined by Barros et al. (2014).

As I mentioned in Sect. 4.1.5.2, the mass of CoRoT-7b is not affected by the choice of model, which attests to the robustness of this result. My mass of 4.73 ± 0.95 M$_\oplus$ is compatible, within uncertainties, with the results found by Queloz et al. (2009), Boisse et al. (2011) and Tuomi et al. (2014). It is within 2-sigma of the masses found by Pont et al. (2010), Hatzes et al. (2011) and Ferraz-Mello et al. (2011).

4.1.5.6 CoRoT-7c

I make a robust detection of CoRoT-7c at an orbital period of 3.70 ± 0.02 days, which is in agreement with previous works that considered planet c. I estimate its mass to be 13.56 ± 1.08 M$_\oplus$ (see Table 4.2). This is in agreement with that given by Boisse et al.

(2011) and Ferraz-Mello et al. (2011). It is just over 2-sigma lower than the mass found by Hatzes et al. (2010), and over 3-sigma greater than the mass calculated by Queloz et al. (2009). It suggests that the harmonic filtering technique employed by Queloz et al. (2009) suppresses the amplitude of the signal at this period. This may be due to the fact that P_c is close to the fifth harmonic of the stellar rotation, $P_{rot}/6 \sim$ 3.9 days (see Fig. 4.4), but Queloz et al. (2009) only model RV variations using the first two harmonics, thus leaving P_c and $P_{rot}/6$ entangled. Ferraz-Mello et al. (2011), who performed a similar analysis to that of Queloz et al. (2009), mention that the proximity of P_c to $P_{rot}/6$ may lead to underestimating the RV amplitude of CoRoT-7c by up to 0.5 m $\cdot$ s^{-1} due to beating between these two frequencies.

I estimated the minimum orbital inclination this planet has to have in order to be transiting. Its radius R_c can be approximated using the formula given by Lissauer et al. (2011):

$$R_c = \left(\frac{M_c}{M_\oplus}\right)^{1/2.06} R_\oplus, \qquad (4.3)$$

where $M_\oplus$ and $R_\oplus$ are the mass and radius of the Earth. Using the mass for CoRoT-7c given in the third column of Table 4.2, I find $R_c = 3.54\, R_\oplus$. With this radius, CoRoT-7c would have to have a minimum orbital inclination of 83.7° in order to be passing in front of the stellar disc with respect to the observer.

CoRoT-7b's orbital axis is inclined at 79.0° to the line of sight (preliminary result of Barros et al. 2014). According to Lissauer et al. (2011), over 85 % of observed compact planetary systems containing transiting super-Earths and Neptunes are coplanar within 3°. Planet c is therefore not very likely to transit. Indeed, no transits of this planet are detected in any of the CoRoT runs. Any planets further out from the star with a similar radius or smaller are even less likely to transit.

4.1.5.7 The Magnetic Activity of CoRoT-7

In Model 2, the RMS scatter of the total activity model is 4.86 m $\cdot$ s^{-1} (see Fig. 4.6b). For moderately active host stars such as CoRoT-7, the activity contribution largely dominates the reflex motion induced by a closely orbiting super-Earth.

The RMS scatter of ΔRV_{rot} and ΔRV_{conv} are 0.46 and 1.82 m $\cdot$ s^{-1}, respectively. The smaller impact of the surface brightness inhomogeneities on the RV variations could be due to the small $v \sin i$ of the star (Bruntt et al. 2010), because the amplitude of these variations scales approximately with $v \sin i$ (Desort et al. 2007). This suggests that for slowly rotating stars such as CoRoT-7, the suppression of convective blueshift is the dominant contributor to the activity-modulated RV signal, rather than the rotational Doppler shift of the flux blocked by starspots. This corroborates the findings of Meunier et al. (2010) and Lagrange et al. (2010), who showed that the suppression of convective blueshift is the dominant source of activity-induced RV variations on the Sun, which is also a slowly rotating star (see discussions back in Chap. 2).

I use a Gaussian process to absorb correlated residuals due to other physical phenomena occurring on timescales of order of the stellar rotation period. In the case of CoRoT-7, these combined signatures have an RMS of 3.95 m · s^{-1}, suggesting that there are other processes than those modelled by the FF' method at play.

4.1.6 Summary

The CoRoT-7 system was re-observed in 2012 with the CoRoT satellite and the HARPS spectrograph simultaneously. These observations allowed me to apply the FF' method of Aigrain et al. (2012) to model the RV variations produced by the magnetic activity of CoRoT-7. If I only use the FF' method to model the activity, I find correlated noise in the RV residuals which cannot be accounted for by a set of Keplerian planetary signals. This indicates that some activity-related noise is still present. Indeed, as previously mentioned in Sect. 2.2.5 in Chap. 2, the FF' method does not account for all phenomena such as the effect of limb-brightened facular emission on the cross-correlation function profile, photospheric inflows towards active regions, or faculae that are not spatially associated with starspot groups. Furthermore, some longitudinal spot distributions have almost no photometric signature (see Sect. 2.1.5.3). To model this low-frequency stellar signal, I use a Gaussian process with a quasi-periodic covariance function that has the same frequency structure as the lightcurve (see Chap. 3).

I run an MCMC simulation and use Bayesian model selection to determine the number of planets in this system and estimate their masses. I find that the transiting super-Earth CoRoT-7b has a mass of $4.73 \pm 0.95\,M_\oplus$. Using the planet radius estimated by Bruntt et al. (2010), CoRoT-7b has a density of $(6.61 \pm 1.72)(R_p/1.58\,R_\oplus)^{-3}$ g · cm^{-3}, which is compatible with a rocky composition. I confirm the presence of CoRoT-7c, which has a mass of $13.56 \pm 1.08\,M_\oplus$. My findings agree with the analyses made by Barros et al. 2014 and Tuomi et al. (2014).

I search for evidence of an additional planetary companion at a period of 9 days, as proposed by Hatzes et al. (2010) following an analysis of the 2008–2009 RV dataset. While the Lomb–Scargle periodogram of the 2012 RVs displays a strong peak in the 6–10 days range, I find that this signal is more likely to be associated with the second harmonic of the stellar rotation at ∼7.9 days.

In CoRoT-7, the RV modulation induced by stellar activity dominates the total RV signal despite the close-in orbit of (at least) one super-Earth and one sub-Neptune-mass planet. Understanding the effects of stellar activity on RV observations is therefore crucial to improve our ability to detect low-mass planets and obtain a precise measure of their mass.

4.2 Kepler-78

Artist impression of Kepler-78b: "an infernal Earth".
Credit: Jasiek Krzysztofiak, Nature.

4.2 Kepler-78

Kepler-78 is a very active star, and would have never been selected for an RV campaign were it not for the discovery of an Earth-size planet crossing its disc every 8.5 h. It was found around the time at which I finished writing my MCMC code with Gaussian processes, so I decided to give it a go. It turns out that a Gaussian process trained on the lightcurve is very effective at modelling activity-driven RV variations for this kind of system.

4.2.1 History of the System

In 2013, Sanchis-Ojeda et al. (2013) reported on the discovery of a transiting short period Earth-size planet around Kepler-78. At the time, this was one of the first planets found with an orbital period of less than 1 day, and it was one of the smallest planets ever discovered. The main characteristics of the star and the transit parameters of Kepler-78b found by Sanchis-Ojeda et al. (2013) are detailed in Table 4.3.

Shortly after this announcement, the star was observed intensively with HARPS-N and HIRES in order to measure the mass of the planet. This was made tricky due to the high levels of activity of the host star. Its full *Kepler* lightcurve, shown in Fig. 4.10 displays peak-to-peak variations of about 10 mmag. By my rule of thumb (acquired from my experience with CoRoT-7, Kepler-10 and results by Aigrain et al. (2012)—see Chap. 2), this translates into activity-induced variations of about $20\,\mathrm{m \cdot s^{-1}}$ peak to peak, and indeed this is what we see in the HIRES and HARPS-N RV observations (see Fig. 4.9).

Table 4.3 Stellar parameters and transit parameters of Kepler-78b (from Sanchis-Ojeda et al. 2013), adopted in my analysis

Kepler-78	
Mass	$0.81 \pm 0.05\, M_\odot$
Radius	$0.74^{+0.10}_{-0.08}\, R_\odot$
Age	750 ± 150 Gyr
Projected rotation, $v \sin i$	2.4 ± 0.5 km $\cdot$ s^{-1}
Kepler-78b	
Orbital period	$0.35500744 \pm 0.00000006$ day
Mid-transit time	$2454953.95995 \pm 0.00015$ BJD
Orbital inclination	79^{+9}_{-14} deg
Radius	$1.16^{+0.19}_{-0.14}\, R_\oplus$
Mass	$1.86^{+0.38}_{-0.25}\, M_\oplus$

They derived the star's age based on its rotation period and mass, using the formula found by Schlaufman (2010) (this age is compatible with the star's projected rotation). The mass of Kepler-78b is that determined by Pepe et al. (2013)

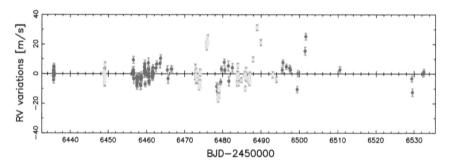

Fig. 4.9 Kepler-78 RV observations by HARPS (*blue points*) and HIRES (*green points*)

Pepe et al. (2013) reported on the HARPS-N observations. In order to determine the mass of the planet, they applied the nightly offsets method of Hatzes et al. (2011), originally developed to measure the mass of CoRoT-7b (see Sects. 2.2.2 and 4.1.1). This technique relies on the stellar activity timescales (the stellar rotation period of about 12 days and its main harmonics) being much longer than the planet orbital period (about 8.5 h). Over the span of a single night, all the variations in RV can be attributed to the planet's motion. It is therefore possible to treat the stellar activity signal as a nightly constant. Using this technique, Pepe et al. (2013) recover a semi-amplitude for Kepler-78b $K_b = 1.96 \pm 0.32$ m $\cdot$ s^{-1}.

Howard et al. (2013) present the analysis of the HIRES RV data. They model the activity-induced RV variations as a sum of Fourier components with periods equal to the stellar rotation period and its first two harmonics (they show that the power at higher harmonics is negligible). This worked well since the activity signals are

4.2 Kepler-78

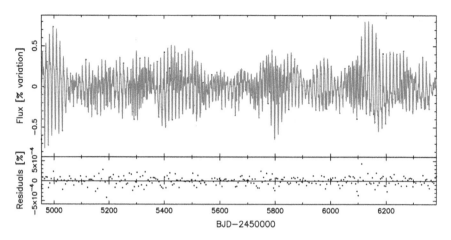

Fig. 4.10 Kepler-78 binned lightcurve, fitted with a Gaussian process (quasi-periodic covariance function)

strongly modulated by the stellar rotation. With this technique, the semi-amplitude obtained for Kepler-78b is $K_b = 1.66 \pm 0.40 \, \text{m} \cdot \text{s}^{-1}$. The semi-amplitudes obtained through both analyses of the two independent datasets are in good agreement.

Following the analyses by Pepe et al. (2013) and Howard et al. (2013), Grunblatt et al. (2015) took a step further and combined the two RV datasets together in order to make a more precise mass determination. The model for activity-induced RV variations is based on a Gaussian process with a quasi-periodic covariance function, trained on the lightcurve in order to extract its frequency structure. We tested a variety of models, including quasi-periodic and square exponential covariance functions, additional white noise parameters and combinations thereof. The two spectrographs have a different wavelength coverage, and the active regions leading to RV variations may produce different amplitudes at different wavelengths, so we also tried modelling the activity signals with a separate Gaussian process for each RV dataset. We compared models in a qualitative way rather than doing a full Bayesian model selection analysis, which we deemed unnecessary at the time. We found that modelling the activity-driven RV variations as two separate Gaussian processes with separate η_1 but the same η_2, η_3 and η_4 hyperparameters, with the addition of two separate white noise terms provided the best fit. The results of this analysis are given in the second column of Table 4.4. We determine the planet mass to a 6.5-sigma precision, an improvement of 2.5-sigma over the value of Howard et al. (2013). Our mass value is in agreement with those of Pepe et al. (2013) and Howard et al. (2013).

The analysis I present here is much simpler: I only use a single Gaussian process for both datasets, with no additional white noise parameter. I will show that both analyses are in agreement. A Gaussian process on its own is effective at modelling activity-induced RV variations reliably, even for a star as active as Kepler-78.

Table 4.4 Outcome of my model, which consists of a Gaussian process with a quasi-periodic covariance function, one Keplerian circular orbit and one zero offset for each RV dataset, compared with the model applied by Grunblatt et al. (2015), consisting of two separate Gaussian processes, one Keplerian circular orbit, two RV offsets and two additional white noise terms (σ)

	My model	Grunblatt et al.
Planet b		
P (days)	$0.35500744 \pm 0.00000006$	
t_0 (BJD—2450000)	$2454953.95995 \pm 0.00015$	
K (m · s^{-1})	1.87 ± 0.19	1.86 ± 0.25
e	0 (fixed)	0 (fixed)
m ($M_\oplus$)	1.76 ± 0.18	$1.87^{+0.27}_{-0.26}$
ρ (g · cm^{-3})	$6.2^{+1.8}_{-1.4}$	$6.0^{+1.9}_{-1.4}$
a (AU)	0.009 ± 0.001	–
Gaussian process for stellar activity		
θ_1 (m · s^{-1})	8.78 ± 1.11	0
$\theta_{1,\text{HARPN}}$ (m · s^{-1})	0	$5.6^{+2.0}_{-1.3}$
$\theta_{1,\text{Keck}}$ (m · s^{-1})	0	$11.6^{+3.7}_{-2.5}$
θ_2 (days)	17 ± 1	$26.1^{+19.8}_{-11}$
θ_3 (days)	12.74 ± 0.06	$13.12^{+0.14}_{-0.12}$
θ_4	0.47 ± 0.05	$0.28^{+0.05}_{-0.04}$
Additional white noise		
σ_{HARPN} (m · s^{-1})	0	$1.1^{+0.4}_{-0.5}$
σ_{Keck} (m · s^{-1})	0	$2.1^{+0.3}_{-0.3}$
Constant RV offsets		
$RV_{0,\text{HARPN}}$ (m · s^{-1})	2.5 ± 3.3	–
$RV_{0,\text{Keck}}$ (m · s^{-1})	-1.0 ± 3.4	–

The quantities marked as '–' in the second column are part of their model, but their values were not listed in the paper

4.2.2 Observations

4.2.2.1 Spectroscopy

The HARPS-N RV campaign spans 2013 May 23-August 28, with 112 observations. I discarded one observation at 24556435.724 BJD as its very low signal-to-noise ratio clearly indicates that it was taken during bad weather. The HIRES campaign, from 2013 June 05 to July 20 overlaps this period, and contains 84 observations. The RVs are shown in Fig. 4.9. The data from both campaigns can be found in Pepe et al. (2013) and Howard et al. (2013).

4.2.2.2 Photometry

Kepler-78 was observed by the *Kepler* satellite at long cadence. Figure 4.10 shows the lightcurve (transits removed) of all quarters concatenated together.

4.2.3 MCMC Analysis

4.2.3.1 RV Model

The orbit of Kepler-78b is modelled as a Keplerian signal. I model the stellar activity RV variations in both RV datasets using a single Gaussian process with a quasi-periodic covariance function trained on the off-transit lightcurve. My final model is as follows:

$$\Delta RV_{\text{tot}}(t_i) = RV_{0,\text{Keck}} + RV_{0,\text{HARPN}} + \Delta RV_{\text{rumble}}(t_i, \theta_1)$$
$$+ \cos(\nu_b(t_i, t_{\text{peri}_b}, P_b) + \omega_b) + e_b \cos(\omega_b)], \quad (4.4)$$

where $RV_{0\text{Keck}}$ and $RV_{0\text{HARPN}}$ are constant offsets for each of the datasets. The period of the orbit of Kepler-78 is given by P_b, and its semi-amplitude is K_b. $\nu_b(t_i, t_{\text{peri}_b})$ is the true anomaly of the planet at time t_i, and t_{peri_b} is the time of periastron. I fix the eccentricity to zero, since with an orbital period of 8.5 h it is reasonable to assume that the planet will be tidally locked to its star.

The period and phase of the planet's orbit are given Gaussian priors centred at the values determined though the photometric analysis of Sanchis-Ojeda et al. (2013), and with a sigma equal to the corresponding error bars of the photometry results.

I determine the parameters of my RV model following my usual MCMC procedure that I described in the previous chapter.

4.2.3.2 Gaussian Process

I choose a quasi-periodic covariance function of the form:

$$k(t, t') = \eta_1^2 \cdot \exp\left(-\frac{(t-t')^2}{2\eta_2^2} - \frac{2\sin^2\left(\frac{\pi(t-t')}{\eta_3}\right)}{\eta_4^2}\right). \quad (4.5)$$

In order to determine the best values of the hyperparameters η, I train the Gaussian process on half the lightcurve, sampled at every 100th point. The resultant lightcurve had 268 points, thus allowing me to compute the covariance matrix (of size 268 × 268) in reasonable time. The sampling corresponds to one point roughly every 2 days, which gives about 6 points per rotation period. Selecting only half the lightcurve still provides me with plenty of rotation cycles in order to estimate the evolution timescale of active regions.

I assume Jeffreys priors for the two timescales η_2 (active-region evolution) and η_3 (rotation period). I also constrain the smoothing factor η_4 to remain between 0 and 1 in order to prevent it from interfering with the evolution timescale. For example, high frequency variations could be accounted for with either a very high value of η_4 or a very small η_2. Constraining η_4 helps avoid this "degeneracy". The best hyperparameter values, determined through the MCMC procedure described in Sect. 3.1.6, are as follows:

1. Amplitude $\eta_1 = 0.0024 \pm 0.0001$ flux units. I subtracted the average value of the flux and then divided by this same value so that the numbers were between 0 and 1;
2. Evolution timescale $\eta_2 = 17 \pm 1$ days. It is longer than the recurrence timescale, which is consistent with the long-lived spots we can see from the autocorrelation function of the lightcurve, shown in Fig. 4.11;
3. Recurrence timescale $\eta_3 = 12.74 \pm 0.06$ days. This rotation period is in agreement with the value $P_{\text{rot}} = 12.71$ days that I get from an autocorrelation analysis, and $P_{\text{rot}} = 12.5 \pm 1$ days found by Sanchis-Ojeda et al. (2013);
4. Smoothing coefficient $\eta_4 = 0.47 \pm 0.05$.

The activity-driven RVs are modelled with a Gaussian process that has the same quasi-periodic covariance function with a set of hyperparameters θ. I set θ_2, θ_3 and θ_4 equal to η_2, η_3 and η_4, respectively. The amplitude θ_1 of the Gaussian process is kept as a free parameter in the MCMC.

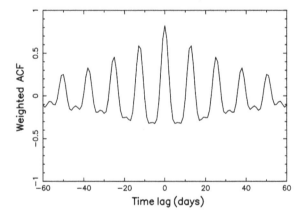

Fig. 4.11 Autocorrelation function of the lightcurve of Kepler-78. It reveals the presence of long-lived active regions, which remain on the stellar disc for several rotations

4.2.4 Results and Discussion

Table 4.4 lists the best-fit parameters I obtain, together with those of Grunblatt et al. (2015). I measure a planet semi-amplitude $K_b = 1.87 \pm 0.19$ m · s^{-1}. The phase-folded orbital signal of Kepler-78b is shown in Fig. 4.12.

My mass determination is consistent with the results found by Grunblatt et al. (2015), Howard et al. (2013) and Pepe et al. (2013). The error bar of my result is slightly smaller than that determined by Grunblatt et al. (2015), but the residuals have an RMS scatter of 2.3 m · s^{-1}, which is slightly higher than the average level of the error bars of 1.92 m · s^{-1} (and the periodogram of the residuals, in panel (d) of Fig. 4.14, shows no significant signals). Grunblatt et al. (2015) use additional white noise terms, which act to increase the error bars of their model in order to bring the RMS of the residuals to the level of the error bars. This additional white noise is likely to come from p-mode oscillations and granulation motions that not have been completely averaged out in each individual RV observation.

All the components of the RV model are shown in Fig. 4.13. Consider the 2 anomalous HARPS-N observations just before day 70; were it not for the presence of similar outliers around days 40 and 50 in the HIRES observations, these points may have been dismissed as outliers affected by instrumental effects or bad weather (this is in fact what Pepe et al. (2013) did). The Gaussian process, however, has no trouble at all accounting for these measurements. This implies that they are compatible with a process that has the same covariance properties, or frequency structure as the lightcurve, and by extension as the magnetic activity behaviour of Kepler-78. The Gaussian process reconciles the two datasets elegantly and effortlessly. Panels (b) and (c) highlight the amplitude difference between the activity-induced variations and the planet orbit. The difference in frequency structure, which is what this analysis relies on, is also highlighted by these plots.

Fig. 4.12 Phase plot of the orbit of Kepler-78b (circular model)

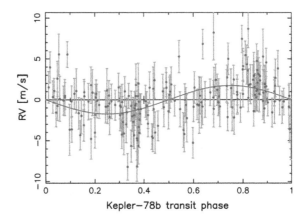

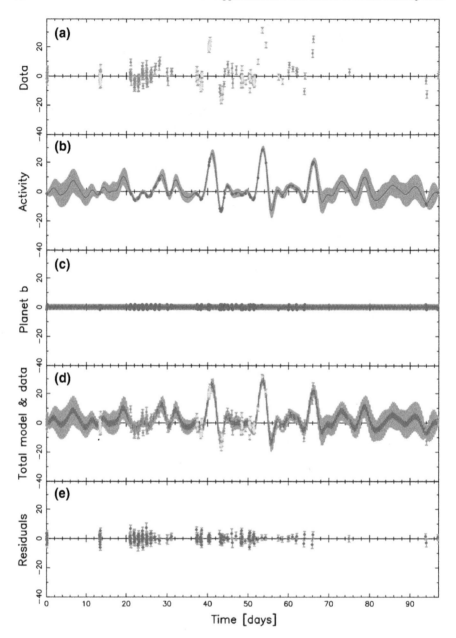

Fig. 4.13 *Panel* **a** the HARPS and HIRES observations, after subtracting the RV offsets for each dataset; *Panel* **b** Gaussian process; *Panel* **c** orbit of Kepler-78b; *Panel* **d** total model (*red*), overlaid on top of the data (*blue points*). *Panel* **e** residuals obtained after subtracting the model from the observations. All RVs are in m · s^{-1}

4.2 Kepler-78

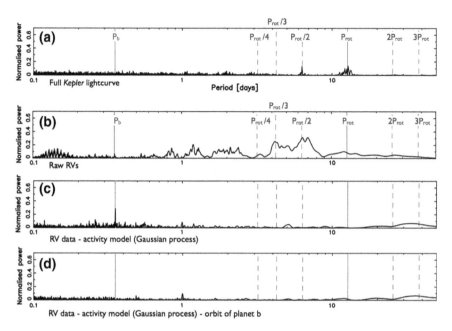

Fig. 4.14 Lomb–Scargle periodograms of: **a** the full *Kepler* lightcurve; **b** the raw RV observations (both datasets); **c** the RV data, from which the activity model has been subtracted, revealing a strong peak at the orbital period of Kepler-78b; **d** same as (**c**), with the orbit of planet b removed

4.2.4.1 The Magnetic Activity of Kepler-78

Figure 4.14 shows Lomb–Scargle periodograms of the *Kepler* lightcurve (panel (a)) and the combined HARPS and HIRES RV data (panel (b)). I computed periodograms on the two RV datasets merged together, assuming a zero RV offset between the two datasets. This is a reasonable assumption, given the estimates of $RV_{0,\mathrm{HARPN}}$ and $RV_{0,\mathrm{Keck}}$ listed in Table 4.4. Although these periodograms are therefore not fully rigorous, they still provide a valuable insight on the frequency structure of the various contributions of my RV model.

The periodogram of the lightcurve, in panel (a) displays strong peaks at P_{rot} and at $P_{\mathrm{rot}}/2$. The periodogram of the raw RVs, in panel (b), is dominated by peaks at $P_{\mathrm{rot}}/2$ and $P_{\mathrm{rot}}/3$; we also see some power at P_{rot} and $P_{\mathrm{rot}}/4$. There is a hint of a peak at the planet's orbital period P_b, but we wouldn't be able to tell the presence of a planet. Once I subtract the Gaussian process, however, the orbit of Kepler-78b becomes clear: the Gaussian process has absorbed the activity signal so successfully that the RV signal due to the planet is detected unambiguously, even without the prior knowledge provided by the *Kepler* transits.

4.2.5 Summary

Following the discovery of the transiting Earth-size planet Kepler-78b, the system was observed with both the HARPS and HIRES spectrographs. I combined these two RV datasets together and used a Gaussian process trained on the lightcurve to model the activity-induced RV variations, which dominate the total RV variations and reach amplitudes of up to 20 m · s^{-1}. I find that the Gaussian process is reliable and effective at accounting for activity-induced signals and allows me to determine a mass for Kepler-78b which is consistent with previous estimates made by Howard et al. (2013), Pepe et al. (2013) and Grunblatt et al. (2015). The precision of my mass determination is slightly better than that of Grunblatt et al. (2015), who analysed the same combined dataset with a model consisting of two separate Gaussian processes and two white noise terms.

4.3 Kepler-10

Kepler-10 could not be more different to Kepler-78. Due to its old age, it is a very quiet star, making it an ideal target for RV follow-up. It is so well-behaved that Dumusque et al. (2014) determined the masses of Kepler-10b and c to excellent precision without the use of any sophisticated activity model—my Bayesian model comparison ruled out the use of a Gaussian process over a simple white noise term by a factor of 10^{16}!

I can still learn valuable lessons from such a system. In this section, I show that my Gaussian process model leaves the planet orbits untouched, allowing me to make an honest determination of their masses.

4.3.1 History of the System

A few months after the discovery of CoRoT-7b, the *Kepler* team announced the detection of several more transiting super-Earths (Borucki et al. 2011). Amongst them, Kepler-10b was the smallest transiting planet yet discovered (Batalha et al. 2011), with a radius of just 1.4 $R_\oplus$. A second planet candidate with an orbital period of about 45 days was also identified, but was not validated by BLENDER (Torres et al. 2011) as scenarios of false positive detections remained too likely with the data available at the time. Follow-up RV observations of Kepler-10 with Keck/HIRES were carried out in order to determine the mass of Kepler-10b (Batalha et al. 2011). Only 40 measurements were obtained, spread over just under a year. These observations yielded a mass with a precision of less than 3-sigma (see Table 4.5) for Kepler-10b. The orbit of Kepler-10c was not detected in the RV measurements, which meant it was only possible to place an upper limit to the mass of this potential additional planetary

4.3 Kepler-10

companion. Further transit observations were later acquired with the *Spitzer* Space Telescope, allowing Fressin et al. (2011) to perform a new BLENDER analysis and validate this second candidate as a small Neptune with a 2.2 $R_\oplus$ radius.

The discovery of such an exciting planetary system prompted Fogtmann-Schulz et al. (2014) to carry out an asteroseismic analysis of the star's physical parameters, using 29 months of *Kepler* photometry instead of only the first 5 months of the mission, as had been done for the discovery paper. The radius, mass and age of Kepler-10 determined by Fogtmann-Schulz et al. (2014) are listed in Table 4.5. Kepler-10 was found to be over 10 Gyr old, which made it the oldest known star to host rocky planets! This also meant that it should be slowly rotating and magnetically quiet, and indeed, its *Kepler* lightcurve, shown in Fig. 4.16, displays almost no variability.

"Kepler-10c, the Godzilla of Earths!" *Term coined by Prof. Dimitar Sasselov*
Artist impression of the Kepler-10 planetary system. *Credit: David Aguilar, Harvard-Smithsonian Center for Astrophysics.*

Table 4.5 Stellar parameters (from Fogtmann-Schulz et al. 2014) and transit parameters of Kepler-10b and c (from Batalha et al. 2011 and Fressin et al. 2011, respectively), adopted in my analysis

Kepler-10	
Mass	$0.913 \pm 0.022\ M_\odot$
Radius	$1.065 \pm 0.008\ R_\odot$
Age	10.50 ± 1.33 Gyr
Projected rotation, $v \sin i$	0.5 ± 0.5 km $\cdot$ s^{-1}
Kepler-10b	
Orbital period	$0.837495^{+0.000004}_{-0.000005}$ day
Mid-transit time	$2454964.57375^{+0.00060}_{-0.00082}$ HJD
Orbital inclination	$84.4^{+1.1}_{-1.6}$ deg
Radius	$1.416^{+0.033}_{-0.036}\ R_\oplus$
Mass	$4.56^{+1.17}_{-1.29}\ M_\oplus$
Kepler-10c	
Orbital period	$45.29485^{+0.00065}_{-0.00076}$ days
Mid-transit time	$2454971.6761^{+0.0020}_{-0.0023}$ HJD
Orbital inclination	$89.65^{+0.09}_{-0.12}$ deg
Radius	$2.227^{+0.052}_{-0.057}\ R_\oplus$

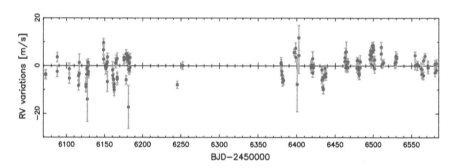

Fig. 4.15 RV variations of Kepler-10 measured with HARPS-N

4.3.2 Observations

4.3.2.1 HARPS-N Spectroscopy

Kepler-10 seemed like a target of choice for RV follow-up, so the HARPS-N team decided to observe Kepler-10 twice per night over several months. The results of this campaign were reported by Dumusque et al. (2014). A few measurements were discarded from the analysis of Dumusque et al. for the following reasons:

4.3 Kepler-10

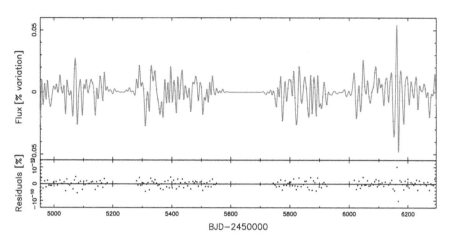

Fig. 4.16 *Upper panel* Selected parts of the (binned) Kepler-10 *Kepler* lightcurve that I used to compute the autocorrelation period (45 days), with my photometric fit overplotted as the *blue curve*. *Lower panel* Residuals of the fit

- Measurements that had a signal-to-noise ratio (at 550 nm) lower than 10 (this was the case for 4 observations);
- All stars observed on the night of 18 October 2013 show an RV offset of 10 m · s^{-1} or more, so we removed the 2 observations taken on this night;
- The original HARPS-N CCD suffered a partial failure in September 2012, and was operated using only the red half of the CCD until a replacement chip was procured and installed in November 2012. We took measurements with only half of the chip for a few nights until the CCD was replaced. This means that the RVs were derived with fewer (and different) spectral lines, so we decided to discard the 5 observations concerned.

This left us with 148 observations, shown in Fig. 4.15. The data are available in Dumusque et al. (2014).

4.3.2.2 *Kepler* photometry

The *Kepler* spacecraft observed Kepler-10 with a 1-minute cadence up to Quarter 14 of the mission (Fogtmann-Schulz et al. 2014). Figure 4.17a shows all the *Kepler* quarters, which I concatenated together by fitting a constant for each—this is a rough procedure but works well (see Chap. 2, Sect. 2.3).

Determination of the stellar rotation period I computed the autocorrelation of the full lightcurve, shown in Fig. 4.17b (see Sect. 2.3.2.2). It yields a rotation period of 45 days. It is apparent, however, that the beginning and end of several quarters display unexpected wiggles that look more like instrument systematics than stellar activity; they are likely to affect our estimate of the rotation period. We (Dumusque

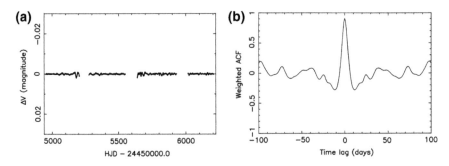

Fig. 4.17 *Panel* **a** the full concatenated PDC-MAP *Kepler* lightcurve, in which instrumental "wiggles" are clearly present; *panel* **b** its autocorrelation function, which indicates a rotation period at 45 days, but shows little structure otherwise. Compare this plot with other similar ones in Fig. 2.11 of Chap. 2 to see just how quiet Kepler-10 is!

et al.) computed the autocorrelation of single quarters, with the wiggles cut out, and consistently arrived at periods between 15 and 16 days. This is significantly different to the 45-day period determined from the concatenated lightcurve. It is also incompatible with the old age of Kepler-10, which points towards a rotation period of at least 22 days (Dumusque et al. 2014). The star's projected rotational velocity (see Table 4.5) is very low and consistent with a period of at least 26 days. Kepler-10 seems to be a case where the PDC-MAP data reduction pipeline erased long-term periodic signals; as I explained in Sect. 2.3 back in Chap. 2, this can unfortunately happen.

4.3.3 MCMC Analysis

4.3.3.1 RV Model

Since we (Dumusque et al.) were not able to obtain a precise and reliable estimate of the rotation period of Kepler-10, we could not justify using a Gaussian process with a quasi-periodic covariance function to account for noise modulated by the rotation of Kepler-10 in the final paper. Besides, the star is very quiet and the RV observations we see in Fig. 4.15 display an RMS scatter of just over $4 \text{ m} \cdot \text{s}^{-1}$, which indicates that any activity-induced variations will be very small and unlikely to significantly affect our planetary mass measurements, and that using a Gaussian process would be excessive (this was confirmed by a Bayesian model comparison which yielded a Bayes' factor of 10^{16} in favour of a white noise term over a Gaussian process). The MCMC analysis presented in Dumusque et al. (2014) accounts for any such variations with a constant white noise term, added in quadrature to the error bars, commonly referred to as a "jitter" term. I did run my code with this model, and the results are included in Sect. 4.5 of Dumusque et al. (2014). Here, I prefer to show the results I obtained with a model comprising a Gaussian process. We shall see that the final mass determinations are compatible with the ones obtained with the model of Dumusque et al. (2014), which attests that the Gaussian process does not absorb the original signals of the two planets.

I ran my MCMC code for a model consisting of two Keplerian orbits, two zero RV offsets in order to account for any changes incurred by the replacement of the CCD, and a Gaussian process to model activity-induced signals, governed by a quasi-periodic covariance function.

As starting points to my MCMC simulation, I adopted the K amplitudes found by the preliminary analyses done in Sects. 4.3 and 4.4 of Dumusque et al. (2014). The orbits of the two planets were constrained by applying Gaussian priors on the orbital period and epochs of transit found by previous photometric analyses, presented in Table 4.5. The usual priors discussed in Sect. 3.2.4 were applied for all other parameters.

4.3.3.2 Gaussian Process

Based on my previous experience of modelling activity-induced RV variations with a Gaussian process, I assumed a quasi-periodic covariance function. Instead of training the Gaussian process on the lightcurve to determine the hyperparameters of the covariance function, I assumed the following hyperparameter values:

- Amplitude: determined via the MCMC procedure that I carried out;
- Recurrence timescale (stellar rotation period): according to the investigations carried out by Dumusque et al. (2014), it is likely to be at least 22 days, and according to the autocorrelation function of the lightcurve in Fig. 4.17, it is likely to be around 45 days. Looking at the periodogram of the RVs in panel (a) of Fig. 4.18, there is a peak at 52 days with clear harmonics at $P/2$, $P/3$ and $P/4$ (see red full and dashed lines). It therefore seems reasonable to assume that this is the stellar rotation period;
- Evolution timescale: I assumed this to be half the rotation period, i.e. 26 days;
- Smoothing parameter: 0.5.

4.3.4 Results and Discussion

4.3.4.1 Selection of the Best Model

I found that a 2-planet model with fixed circular orbits is preferred over a model with free eccentricities by a factor of 2047 (according to Jeffreys (1961), a Bayes' factor over 150 indicates strong evidence). When the eccentricities are let free, I obtain $e_b = 0.002 \pm 0.002$ and $e_c = 0.002 \pm 0.05$, which suggests that both orbits

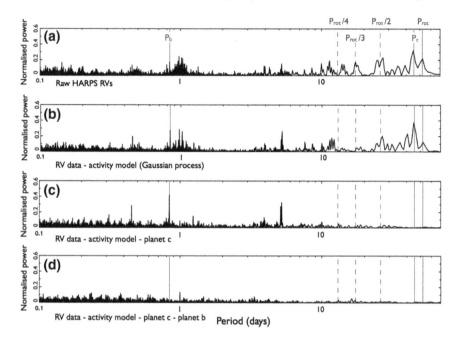

Fig. 4.18 Lomb–Scargle periodograms of: **a** the raw RV observations; **b** the RV data, from which the activity RV model has been subtracted; **c** same as (**b**), with the RV signal of planet c also removed; **d** finally, the RV contribution of planet b is also removed

are compatible with circular orbits. Furthermore, I see no significant difference in the planet masses or the RV residuals.

4.3.4.2 Best Model

The results for a 2-planet model with forced circular orbits are listed in Table 4.6. The K amplitudes found for both planets in agreement, within 1-sigma, with the main MCMC analysis presented in Dumusque et al. (2014). The uncertainty on my mass determination of Kepler-10b is smaller than that of Dumusque et al. (2014), but in the case of Kepler-10c this goes the other way around.

The Gaussian process framework is more flexible than a white noise term. My intuition tells me that the Gaussian process will allow me to determine the masses of planets at different orbital periods with varying levels of uncertainty, depending on their "temporal proximity", i.e. their degree of overlap with the frequency structure of the Gaussian process. If this were the case, the results could be interpreted as follows:

- The orbital period of Kepler-10b (0.85 day) is very distinct from the frequency structure of the Gaussian process (52 days and harmonics thereof). The Gaussian

4.3 Kepler-10

Table 4.6 Outcome of my model, which consists of a Gaussian process with a quasi-periodic covariance function, 2 planet orbits with eccentricities fixed to 0 and two zero offsets (to account for the CCD replacement), compared with the results of the model applied by Dumusque et al. (2014), in which the Gaussian process is replaced by a white noise term

	My model	Dumusque et al. (2014)
Kepler-10b		
P (days)	0.8374907(2)	0.8374907(2)
t_0 (BJD—2450000)	5034.0868(2)	5034.0868(2)
K (m · s^{-1})	2.37 ± 0.23	2.38 ± 0.34
e	0 (fixed)	0 (fixed)
m ($M_\oplus$)	3.31 ± 0.32	3.33 ± 0.49
ρ (g · cm^{-3})	$6.4^{+1.1}_{-0.7}$	5.8 ± 0.8
a (AU)	0.016(1)	–
Kepler-10c		
P (days)	45.29429(4)	45.29430(4)
t_0 (BJD—2450000)	5062.2664(4)	5062.26648(8)
K (m · s^{-1})	3.09 ± 0.69	3.25 ± 0.36
e	0 (fixed)	0 (fixed)
m ($M_\oplus$)	16.2 ± 3.6	17.2 ± 1.9
ρ (g · cm^{-3})	8.1 ± 1.8	7.1 ± 1.0
a (AU)	0.24(1)	–
Additional noise		
θ_1 (m · s^{-1})	2.37 ± 0.34	0
σ_s (m · s^{-1})	0	$2.45^{+0.23}_{-0.21}$

The numbers in brackets represent the uncertainty in the last digit of the value

process is unlikely to interfere with the orbit of Kepler-10b. The model is therefore able to unambiguously identify this signal, yielding a precise mass determination.
- The orbital period of Kepler-10c, on the other hand, is much closer to the recurrence timescale governing the structure of the Gaussian process (35 days—see the first two periodograms of Fig. 4.18). In this region of parameter space, it is therefore more tricky to isolate the orbital signature of the planet. The uncertainty on the mass determination increases in order to reflect this.
- In comparison, a white noise term would provide a constant level of uncertainty regardless of the stellar activity and orbital timescales.

It would be of utmost interest to see whether this is indeed the case, and further investigation is required. I discuss a possible future project to tackle this at the end of this chapter.

Figure 4.19 shows each component of the total RV model. We cannot see the variations of Kepler-10b very clearly because its orbital period is very short, but from this plot we can get a sense of the relative amplitudes of the Gaussian process and the two planets, and see over which timescales each one of them is important.

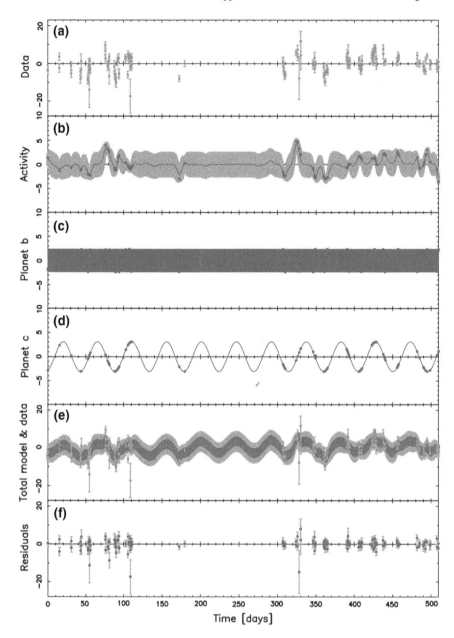

Fig. 4.19 *Panel* **a** HARPS observations, after subtracting the star's systemic velocity RV_0; *Panel* **b** Gaussian process activity model; *Panel* **c** orbit of Kepler-10b; *Panel* **d** orbit of Kepler-10c; *Panel* **e** total model (*red*), overlaid on top of the data (*blue points*). *Panel* **f** residuals obtained after subtracting the model from the observations. Note that the scale on the *y*-axis in panels (**b**), (**c**) and (**d**) differ from the other panels. All RVs are in m · s^{-1}

4.3 Kepler-10

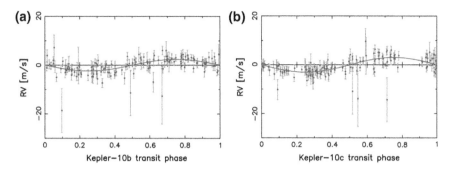

Fig. 4.20 *Panel* **a** Phase plot of the orbit of planet b (circular model). *Panel* **b** Phase plot of the orbit of planet c (circular model)

Figure 4.18 shows the periodogram of the HARPS-N RV data in panel (a), and the effect of removing each planet orbit one at a time. Removing the activity RV model in panel (b) reveals the two planets. When the orbit of Kepler-10b is removed in panel (d), we see that the peak at 0.8 days and its 1-day alias at 4.8 days both disappear. The RV residuals remaining after subtracting my model from the observations have an RMS scatter of 3 m $\cdot$ s^{-1}, which is about 1 m $\cdot$ s^{-1} greater than the average size of the error bars. On the last panel of Fig. 4.13, however, we can see that the majority of the residuals are close to zero, while a few isolated points are very far off (they also have larger error bars). This additional 1 m $\cdot$ s^{-1} is therefore likely to be caused by these few outliers. As shown in panel (d) of Fig. 4.18, there are no obvious peaks in the generalised Lomb–Scargle periodogram of the residuals.

I show the phase-folded plots of the two planets in Fig. 4.20.

4.3.5 Summary

Following the discovery of two transiting planets, one of them an Earth-size planet, the Kepler-10 system was observed intensively with the HARPS-N spectrograph in order to determine the masses of the planets. Kepler-10 is very old and quiet so a complex activity model was not required; nevertheless, I wished to test whether a Gaussian process still works when it is not needed. This system proved to be a double challenge when it was established that its *Kepler* lightcurve cannot be trusted to reveal the magnetic activity frequency structure of the star; this is the case for a number of lightcurves as cautioned by the *Kepler* Data Release 21 Notes (refer to Sect. 2.3). Based on my previous experience and on existing analyses of this system, I made guesses for the rotation period and lifetime of active regions and went on to run my MCMC simulation to determine the best-fitting parameters of my RV model.

My planet mass determinations are in agreement, within 1-sigma, with those of Dumusque et al. (2014). The uncertainties found via both methods are different, and I plan to investigate this further. In any case, this analysis shows that the Gaussian process model does not absorb the planetary signals and provides robust mass determinations.

4.4 Summary and Future Plans

4.4.1 Determining the Bulk Densities of Transiting Exoplanets

Determining the mass of a transiting planet allows us to infer its bulk density, since we can measure its radius from the transits in the photometry. This gives us an insight into what the planet is made of, and what its structure might be like. A precision of at least 10% in mass (and 5% in radius) is required to distinguish rocky planets with iron cores from planets made mostly of water (Zeng and Sasselov 2013). This is very challenging, but thanks to *Kepler*, soon TESS, CHEOPS and PLATO, and spectrographs such as TNG/HARPS-North and eventually VLT/ESPRESSO, it is becoming a reality!

If we can obtain this information for a large number of exoplanets it can provide essential clues on the processes that led to the formation of these planetary systems.

Using the radius found by Bruntt et al. (2010), I find that CoRoT-7b is slightly denser than the Earth ($\rho_\oplus = 5.52$ g·cm^{-3}), with $\rho_b = 6.61 \pm 1.72$ g·cm^{-3} (see Table 4.2). Refer to Barros et al. (2014) for a more detailed discussion of the density of CoRoT-7b.

Kepler-10 c has a density of 7.1 ± 1.0 g·cm^{-3} (value of Dumusque et al. 2014), which indicates that the planet is of rocky composition. Based on current theories of planet formation, this was an unexpected discovery, and earned Kepler-10c the name "Godzilla Earth".

I placed CoRoT-7b, Kepler-10b, Kepler-10c and Kepler-78b on a mass-radius diagram alongside other exoplanets for which mass and radius have been measured in Fig. 4.21. According to composition models by Zeng and Sasselov (2013), CoRoT-7b, Kepler-10c and Kepler-78b along with Kepler-20b all have the density expected of a rocky planet; we see that they lie along the black "rocky" line of the diagram, despite displaying a range of radii. Kepler-10b is slightly less dense and its bulk density is more consistent with a composition of half-rock, half-ice.

4.4.2 Assessing the Reliability of the Gaussian Process Framework for Exoplanet Mass Determinations

As unveiled in Chap. 3, I have developed a new data analysis tool that can reproduce the effects of stellar activity in RV observations by using a Gaussian process trained on the variations in the stars lightcurve.

4.4 Summary and Future Plans

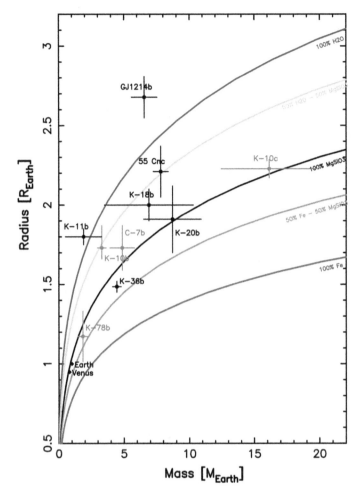

Fig. 4.21 CoRoT-7b, Kepler-10b, Kepler-10c and Kepler-78b on a mass-radius diagram. Earth and Venus are shown as diamond shaped symbols for comparison. Other exoplanets for which the radius and mass are known are also represented. The *solid lines* show mass and radius for planets consisting of (from *top* to *bottom*): pure water, 50 % water and 50 % silicates, pure silicates, 50 % silicates and 50 % iron core, and pure iron, according to the theoretical models of Zeng and Sasselov (2013)

In the present chapter, I reported on applications of my code to three low-mass planetary systems: CoRoT-7, Kepler-78 and Kepler-10. My Gaussian process framework works well for a variety of magnetic activity levels, and it has the potential to become a state of the art tool for exoplanet characterisation in future years. In order to achieve this, my code requires further testing and systematic benchmarking before it is applied in a more automated way to a large number of planetary systems. I propose the following project:

1. Create sets of synthetic RV data and synthetic lightcurves. The synthetic RV datasets would be a combination of one or more planetary orbits (for planets of various masses, orbital periods and orbital eccentricities) and white and red noise, to reproduce instrumental and astrophysical noise. I would design red noise with a quasi-periodic behaviour intended to mimic the effects of stellar activity on RV observations, which are strongly modulated by the stellar rotation period, and depend on the growth and decay of active regions on the stellar surface. The synthetic lightcurves would consist of a simple Fourier series with decaying amplitudes, with white noise.
2. Apply my code to the synthetic datasets. I would check the results to see if the code can detect the fake injected planet signals. As an extension of the investigations I undertook to assess the existence of CoRoT-7d, I would determine how well the code performs for each synthetic model and identify configurations for which the planetary signals are not fully recovered. In particular, I would test whether my code is capable of detecting planets with orbital periods close to the stellar rotation period or its harmonics. I could produce a plot showing the detectability of planets as a function of "temporal proximity".
3. I could further automatise my code so that I can then easily run it on a large number of stars, for example the HARPS-N database, in order to help us determine the number of observations we need for individual planet systems in order to determine planet masses with a 3- (or 6-) sigma precision, for a given radius and assumed composition.

It would also be interesting to carry out rigorous Bayesian model comparison on the Kepler-78 combined dataset, and possibly for other systems with observations from different spectrographs to find out whether two separate Gaussian processes perform better at modelling activity-induced signals as opposed to only one. This would tell us whether the RV amplitude of variations induced by active regions does change significantly as a function of wavelength, to the extent that we can detect these differences with HARPS and HIRES (or other spectrographs).

4.4.3 Concluding Note

The intrinsic variability of the stars themselves remains the main obstacle to determining the masses of small planets. It is essential that we develop effective and comprehensive data analysis techniques, and that we establish reliable proxies for activity-induced RV signals to be able to extract the planetary signals from stellar variability. In the next chapter, I present the work I have done on the activity-driven RV variations of the Sun, in the aim to break this barrier.

References

Aigrain S, Pont F, Zucker S (2012) Mon Not R Astron Soc 419:3147
Auvergne M et al (2009) Astron Astrophys 506:411
Barros SCC et al (2014) Astron Astrophys 569:74
Batalha NM et al (2011) Astrophys J 729:27
Boisse I, Bouchy F, Hébrard G, Bonfils X, Santos N, Vauclair S (2011) Astron Astrophys 528:A4
Borucki WJ et al (2011) Astrophys J 728:117
Bruntt H et al (2010) Astron Astrophys 519:A51
Chib S, Jeliazkov I (2001) American statistical association portal : marginal likelihood from the metropolis-hastings output. J Am Stat Assoc 96(453):270
Desort M, Lagrange AM, Galland F, Udry S, Mayor M (2007) Astron Astrophys 473:983
Dumusque X et al (2014) Astrophys J 789:154
Dumusque X et al (2012) Nature
Edelson RA, Krolik JH (1988) Astrophys J 333:646
Ferraz-Mello S, Tadeu dos Santos M, Beaugé C, Michtchenko TA, Rodríguez A (2011) Astron Astrophys 531:A161
Fogtmann-Schulz A, Hinrup B, Van Eylen V, Christensen-Dalsgaard J, Kjeldsen H, Silva Aguirre V, Tingley B (2014) ApJ 781:67 arXiv:1311.6336
Fressin F et al (2011) Astrophys J Suppl Ser 197:5
Grunblatt SK, Howard AW, Haywood RD (2015) ApJ 808:127 arXiv:1501.00369
Hatzes AP et al (2010) Astron Astrophys 520:A93
Hatzes AP et al (2011) Astrophys J 743:75
Haywood RD et al (2014) Mon Not R Astron Soc 443(3):2517–2531, 443, 2517
Howard AW et al (2013) Nature 503:381
Hussain G (2002) Astronomische Nachrichten 323:349
Jeffreys SH (1961) The theory of probability. Oxford University Press, Oxford
Kopp G, Lean JL (2011) Geophys Res Lett 38:L01706
Lagrange AM, Desort M, Meunier N (2010) Astron Astrophys 512:A38
Lanza AF, Boisse I, Bouchy F, Bonomo AS, Moutou C (2011) Astron Astrophys 533:A44
Lanza AF et al (2010) Astron Astrophys 520:A53
Lanza AF et al (2009) Astron Astrophys 493:193
Léger A et al (2009) Astron Astrophys 506:287
Lissauer JJ et al (2011) Astrophys J Suppl Ser 197:8
Meunier N, Desort M, Lagrange AM (2010) Astron Astrophys 512:A39
Pepe F et al (2013) Nature 503:377
Pont F, Aigrain S, Zucker S (2010) Mon Not R Astron Soc 411:1953–1962
Queloz D et al (2009) Astronomy and Astrophysics 506:303
Sanchis-Ojeda R, Rappaport S, Winn JN, Levine A, Kotson MC, Latham DW, Buchhave LA (2013) Astrophys J 774:54
Schlaufman KC (2010) Astrophys J 719:602
Schrijver CJ (2002) Astronomische Nachrichten 323:157
Torres G et al (2011) Astrophys J 727:24
Tuomi M, Anglada-Escudé G, Jenkins JS, Jones HRA (2014) MNRAS, pre-print arXiv:1405.2016
Zeng L, Sasselov D (2013) PASP 125:227 arXiv:1301.0818

Chapter 5
An Exploration into the Radial-Velocity Variability of the Sun

The presence of starspots, faculae and granulation on the photosphere of a star induces quasi-periodic signals that can conceal and even mimic the Doppler signature of orbiting planets. This has resulted in several false detections (see Queloz et al. 2001; Bonfils et al. 2007; Huélamo et al. 2008; Boisse et al. 2009, 2011; Gregory 2011; Haywood et al. 2014; Santos et al. 2014; Robertson et al. 2014 and many others). Understanding the RV signatures of stellar activity, in particular those modulated by the stellar rotation, is essential to develop the next generation of more sophisticated activity models and further improve our ability to detect and characterise low-mass planets.

The Sun is the only star surface can be directly resolved at high resolution, and therefore constitutes an excellent test case to explore the physical origin of stellar radial-velocity variability. In this chapter, I present HARPS observations of sunlight scattered off the bright asteroid 4/Vesta, from which I deduced the Sun's activity-driven RV variations. In parallel, the HMI instrument onboard the Solar Dynamics Observatory provided me with simultaneous high spatial resolution magnetograms, dopplergrams, and continuum images of the Sun. I determined the RV modulation arising from the suppression of granular blueshift by magnetically active regions (sunspots and faculae) and the flux imbalance induced by dark spots. I confirm that the inhibition of convection is the dominant source of activity-induced RV variations at play. Finally, I find that the activity-driven RV variations of the Sun are strongly correlated with its full-disc magnetic flux, which could become a useful proxy for activity-related RV noise in future exoplanet searches.

This chapter uses material from, and is based on, Haywood et al., 2016, MNRAS, 457, 3637.

© Springer International Publishing Switzerland 2016
R.D. Haywood, *Radial-velocity Searches for Planets Around Active Stars*,
Springer Theses, DOI 10.1007/978-3-319-41273-3_5

5.1 Previous Studies on the Intrinsic RV Variability of the Sun

The Sun is a unique test case as it is the only star whose surface can be resolved at high resolution, therefore allowing me to directly investigate the impact of magnetic features on RV observations. Early attempts to measure the RV of the integrated solar disc did not provide quantitative results about the individual activity features responsible for RV variability. Jiménez et al. (1986) measured integrated sunlight using a resonant scattering spectrometer and found that the presence of magnetically active regions on the solar disc led to variations of up to 15 m $\cdot$ s^{-1}. They also measured the disc-integrated magnetic flux but didn't find any significant correlation with RV at the time due to insufficient precision. At about the same time, Deming et al. (1987) obtained spectra of integrated sunlight with an uncertainty level below 5 m$\cdot$s^{-1}, enabling them to see the RV signature of supergranulation. The trend they observed over the 2-year period of their observations was consistent with suppression of convective blueshift from active regions on the solar surface. A few years later, Deming and Plymate (1994) confirmed the findings of both Jiménez et al. (1986) and Deming et al. (1987), only with a greater statistical significance. Not all studies were in agreement with each other, however; McMillan et al. (1993) recorded spectra of sunlight scattered off the Moon over a 5-year period and found that any variations due to solar activity were smaller than 4 m $\cdot$ s^{-1}.

More recently, Molaro and Centurión (2010) obtained HARPS spectra of the large and bright asteroid Ceres to construct a wavelength atlas for the Sun. They found that these spectra of scattered sunlight provide precise disc-integrated solar RVs, and proposed using asteroid spectra to calibrate high precision spectrographs used for planet hunting, such as HIRES and HARPS. In parallel, significant discoveries were made towards a precise quantitative understanding of the RV impact of solar surface features. Lagrange et al. (2010) and Meunier et al. (2010) used a catalogue of sunspot numbers and sizes and magnetograms from MDI/SOHO to simulate integrated-Sun spectra over a full solar cycle and deduce the impact of sunspots and networks of faculae on RV variations. Flux blocked by sunspots was found to cause variations of the order of the m $\cdot$ s^{-1} (Lagrange et al. 2010; Makarov et al. 2009), while facular suppression of granular blueshift can lead to variations in RV of up to 8–10 m $\cdot$ s^{-1} (Meunier et al. 2010). In particular, it seems that the suppression of granular blueshift by active regions plays a dominant role (Meunier et al. 2010; Haywood et al. 2014).

Following the launch of the Solar Dynamics Observatory (SDO, Pesnell et al. 2012) in 2010, continuous observations of the solar surface brightness, velocity and magnetic fields have become available with image resolution finer than the photospheric granulation pattern. This allows me to probe the RV variations of the Sun in unprecedented detail. In this chapter, I deduce the activity-driven RV variations of the Sun based on HARPS observations of the bright asteroid Vesta (Sect. 5.2). In parallel, I use high spatial resolution continuum, dopplergram and magnetogram images from the Helioseismic and Magnetic Imager (HMI/SDO, Schou et al. 2012) to model the individual RV contributions from sunspots, faculae and granulation (Sect. 5.3). This

5.1 Previous Studies on the Intrinsic RV Variability of the Sun

allows me to create a model which I test against the HARPS observations (Sect. 5.4). Finally, I compute the disc-averaged magnetic flux and show that it is an excellent proxy for activity-driven RV variations (Sect. 5.5).

5.2 HARPS Observations of Sunlight Scattered Off Vesta

5.2.1 HARPS Spectra

The HARPS spectrograph, mounted on the ESO 3.6 m telescope at La Silla was used to observe sunlight scattered from the bright asteroid 4/Vesta (its average magnitude during the run was 7.6). Two to three measurements per night were made with simultaneous Thorium exposures for a total of 98 observations, spread over 37 nights between 2011 September 29 and December 7. The geometric configuration of the Sun and Vesta relative to the observer is illustrated in Fig. 5.1. At the time of the observations, the Sun was just over three years into its 11-year magnetic cycle; the SDO data confirm that the Sun showed high levels of activity.

The spectra were reprocessed using the HARPS DRS pipeline (Baranne et al. 1996; Lovis and Pepe 2007). Instead of applying a conventional barycentric correction, the wavelength scale of the calibrated spectra was adjusted to correct for the doppler shifts due to the relative motion of the Sun and Vesta, and the relative motion of Vesta and the observer (see Sect. 5.2.3). The FWHM and BIS of the cross-correlation function and $\log R'_{HK}$ index were also derived by the pipeline. The median, minimum and maximum signal to noise ratio of the reprocessed HARPS spectra at central wavelength 556.50 nm are 161.3, 56.3 and 257.0, respectively. Overall, HARPS achieved a precision of 75 ± 25 cm s^{-1}.

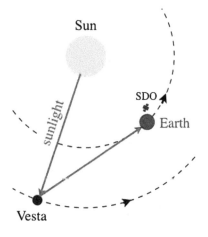

Fig. 5.1 Schematic representation of the Sun, Vesta and Earth configuration during the period of observations (not to scale)

I account for the RV modulation induced by Vesta's rotation in Sect. 5.2.4.1, and investigate sources of intra-night RV variations in Sect. 5.2.4.2. I selected the SDO images in such a way as to compensate for the different viewing points of Vesta and the SDO spacecraft: Vesta was trailing SDO, as shown in Fig. 5.1. This is taken into account in Sect. 5.2.5.

5.2.2 Solar Rest Frame

The data reduction pipeline for HARPS assumes that the target observed is a distant point-like star, and returns its RV relative to the solar system barycenter (RV_{bary}). In order to place the observed RVs of Vesta in the solar rest frame, I perform the following operation:

$$RV = RV_{bary,Earth} + v_{sv} + v_{ve}, \qquad (5.1)$$

where $RV_{bary,Earth}$ is the barycentric RV of the Earth, i.e. the component of the observer's velocity relative to the solar system barycentre, toward the apparent position of Vesta. It can be found in the fits header for each observation. The two components v_{sv} and v_{ve}, retrieved from the JPL HORIZONS database[1] correspond to:

- v_{sv}: the velocity of Vesta relative to the Sun at the instant that light received at Vesta was emitted by the Sun;
- v_{ve}: the velocity of Vesta relative to Earth at the instant that light received by HARPS was emitted at Vesta.

This correction accounts for the RV contribution of all bodies in the solar system and places the Sun in its rest frame.

5.2.3 Relativistic Doppler Effects

The only relativistic corrections made by JPL HORIZONS are for gravitational bending of the light and relativistic aberration due to the motion of the observer (Giorgini, priv. comm.). We therefore must correct for the relativistic doppler shifts incurred by space-time path curvature between the target and the observer. The wavelength correction factor to be applied is given by Lindegren and Dravins (2003) as:

$$\frac{\lambda_e}{\lambda_o} = \frac{\sqrt{1 - \frac{v^2}{c^2}}}{1 + \frac{v \cos \theta_o}{c}}, \qquad (5.2)$$

---

[1] Solar System Dynamics Group, Horizons On-Line Ephemeris System, 4800 Oak Grove Drive, Jet Propulsion Laboratory, Pasadena, CA 91109 USA—Information: http://ssd.jpl.nasa.gov/, Jon.Giorgini@jpl.nasa.gov.

5.2 HARPS Observations of Sunlight Scattered Off Vesta

where λ_e is the wavelength of the light at emission, λ_o is the wavelength that is seen when it reaches the observer, and c is the speed of light. v is the total magnitude of the velocity vector of the observer relative to the emitter. I apply this correction twice:

- The light is emitted by the Sun and received at Vesta. In this case, v is the magnitude of the velocity of Vesta with respect to the Sun, and the radial component $v \cos \theta_o$ is equal to v_{sv} (defined in Sect. 5.2.2).
- Scattered sunlight is emitted from Vesta and received at La Silla. v is the magnitude of the velocity of Vesta with respect to an observer at La Silla, and $v \cos \theta_o$ is v_{ve}.

For both cases, v and can be obtained from the JPL HORIZONS database. All velocities are measured at the flux-weighted mid-exposure times of observation (MJD$_{mid}$_UTC).

The two wavelength correction factors are then multiplied together in order to compute the total relativistic correction factor to be applied to the pixel wavelengths in the HARPS spectra, from which I derive the correct RVs, shown in Fig. 5.2a (see Appendix table of Haywood et al. 2016).

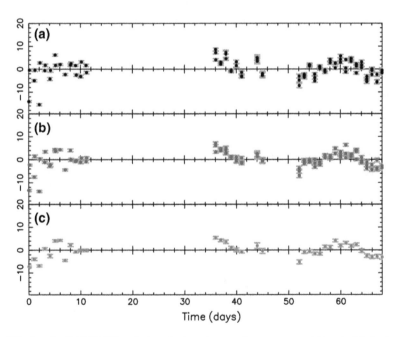

Fig. 5.2 *Panel* **a** HARPS RV variations in the solar rest-frame, corrected for relativistic doppler effects (but not yet corrected for Vesta's axial rotation). *Panel* **b** HARPS RV variations of the Sun as-a-star (after removing the RV contribution of Vesta's axial rotation). *Panel* **c** Nightly binned HARPS RV variations of the Sun as-a-star. All RVs are in ms^{-1}

5.2.4 Sources of Intra-Night RV Variations

5.2.4.1 Vesta's Axial Rotation

Vesta rotates every 5.34 h (Stephenson 1951), so any significant inhomogeneities in its shape or surface albedo will induce an RV modulation. Vesta's shape is close to a spheroid (Thomas et al. 1997), and Lanza and Molaro (2015) found that the RV modulation expected from shape inhomogeneities should not exceed 0.060 m s^{-1}.

Stephenson (1951) presented a photometric study of the asteroid, and suggested that its surface brightness is uneven. He reported brightness variations $\delta m = 0.12$ mag. To make a rough estimate of the amplitude of the RV modulation, I can assume that the brightness variations are due to a single dark equatorial spot on the surface of Vesta, blocking a fraction δf of the flux f. δm and δf are related as follows:

$$\delta m = \frac{2.5 \, d(\ln f)}{\log(e)} \sim 1.08 \, \frac{\delta f}{f}, \tag{5.3}$$

The fractional flux deficit caused by a dark spot can thus be approximated as:

$$\frac{\delta f}{f} \sim \delta m / 1.08 \sim 0.11. \tag{5.4}$$

When the dayside of Vesta is viewed fully illuminated, this spot will give an RV modulation equal to:

$$\Delta RV_{\text{vesta}} = -\frac{\delta f}{f} \, v_{\text{eq}} \, \cos\theta \, \sin\theta, \tag{5.5}$$

where θ is the angle between the spot on the asteroid and our line of sight, and increases from $-\pi/2$ to $+\pi/2$ as it traverses the visible daylight hemisphere. Due to foreshortening, the RV contribution is decreased by a factor $\cos\theta$. The line-of-sight velocity varies with $\sin\theta$. The asteroid's equatorial velocity v_{eq} is given by:

$$v_{\text{eq}} = 2\pi \, \frac{R_{\text{vesta}}}{P_{\text{rot}}}. \tag{5.6}$$

Using a mean radius $R_{\text{vesta}} = 262.7$ km (Russell et al. 2012) and the rotational period $P_{\text{rot}} = 5.34$ h, I obtain $v_{\text{eq}} = 85.8$ m s^{-1}. The maximum RV amplitude of Vesta's rotational modulation, expected at $\theta = \pi/4$ is thus approximately 4.7 m s^{-1}. The RV modulation due to surface brightness inhomogeneity should therefore dominate strongly over shape effects.

I find that this RV contribution is well modelled as a sum of Fourier components modulated by Vesta's rotation period:

$$\Delta RV_{\text{vesta}}(t) = C \cos(2\pi - \lambda(t)) + S \sin(2\pi - \lambda(t)), \tag{5.7}$$

5.2 HARPS Observations of Sunlight Scattered Off Vesta

where $\lambda(t)$ is the apparent planetographic longitude of Vesta at the flux-weighted mid-times of the HARPS observations and can be retrieved via the JPL HORIZONS database (the values of λ are given in the Appendix of Haywood et al. 2016). C and S are scaling parameters, which I determine via an optimal scaling procedure described in Sect. 5.4. Since the phase-folded lightcurve of Vesta shows a double-humped structure (Stephenson 1951), I also tested adding further Fourier terms modulated by the first harmonic of the asteroid's rotation. The improvement to the fit was negligible, so I preferred the simpler model of Eq. 5.7.

Figure 5.2b shows the RV observations obtained after subtracting Vesta's rotational signature. The night-to-night scatter has been reduced, even though much of it remains in the first block of observations; I discuss this in the following section.

5.2.4.2 Solar P-Modes and Granulation

The RV variations in the first part of the HARPS run (nights 0 to 11 in Fig. 5.2) contain some significant scatter, even after accounting for Vesta's rotation. This intra-night scatter does not show in the solar FWHM, BIS or $\log(R'_{HK})$ variations. I investigated the cause of this phenomenon and excluded changes in colour of the asteroid or instrumental effects as a potential source of additional noise. Vesta was very bright (7.6 mag), so I deem the phase and proximity of the Moon unlikely to be responsible for the additional scatter observed.

Solar p-mode oscillations dominate the Sun's power spectrum at a timescale of about 5 min. Most of the RV oscillations induced by p-mode acoustic waves are therefore averaged out within the 15-minute HARPS exposures. Granulation motions result in RV signals of several m s^{-1}, over timescales ranging from about 15 min to several hours. Taking multiple exposures each night and averaging them together (as plotted in panel (c) of Fig. 5.2) can help to significantly reduce granulation-induced RV variations. In addition to this, super-granulation motions commonly take place over timescales of 8 h or longer, and could potentially result in residual white noise from one night to the next. Two different observing strategies were implemented during the HARPS run:

- *First part (nights 0–11):* 2 to 3 observations were made on each night at ∼2-h intervals. Within each night, I see scatter with an amplitude of several m s^{-1} (see panel (b) of Fig. 5.2). I attribute this to granulation motions with a turnover timescale of 2–3 h, that are not averaged well with this observational strategy. When I consider the nightly averages (panel (c)), the scatter is considerably reduced, although some residual noise with an amplitude of ∼3 m s^{-1} remains.
- *Second part (nights 36–68):* 3 consecutive exposures were made on each night. This strategy appears to average out granulation motions very effectively, as little intra-night scatter remains.

The remaining variations, of order 7–10 m s^{-1}, are modulated by the Sun's rotation and are caused by the presence of magnetic surface markers, such as sunspots and faculae. These variations are the primary focus of this chapter, and I model them using SDO/HMI data in Sect. 5.3.

5.2.5 Time Lag Between Vesta and SDO Observations

At the time of the observations, the asteroid Vesta was trailing the SDO spacecraft, which orbits the Earth (see Fig. 5.1). In order to model the solar hemisphere facing Vesta at time t, I used SDO images recorded at $t + \Delta t$, where Δt is proportional to the difference in the Carrington longitudes of the Earth/SDO and Vesta at the time of the HARPS observation. These longitudes can be retrieved from the JPL HORIZONS database. The shortest delay, at the start of the observations was ~ 2.8 days, while at the end of the observations it reached just over 6.5 days (see Appendix table of Haywood et al. 2016). I cannot account for the evolution of the Sun's surface features during this time, and must assume that they remain frozen in this interval. The emergence of sunspots can take place over a few days, but in general large magnetic features (sunspots and networks of faculae) evolve over timescales of weeks rather than days.

5.3 Pixel Statistics from SDO/HMI Images

In the second part of this analysis I aim to determine the RV contribution from granulation, sunspots and facular regions. I used high-resolution full-disc continuum intensity (6000 Å), line-of-sight doppler images and line-of-sight magnetograms from the HMI instrument (Helioseismic and Magnetic Imager) onboard SDO.[2] These were retrieved for the period spanning the HARPS observations of Vesta at times determined by the time lags detailed in Sect. 5.2.5. SDO/HMI images the solar disc at a cadence of 45 sec, with a spatial resolution of 1" using a CCD of 4096×4096 square pixels. I first converted the SDO/HMI images from pixel coordinates to heliographic coordinates, i.e. to a coordinate system centered on the Sun. This coordinate system is fixed with respect to the Sun's surface and rotates in the sidereal frame once every 25.38 days, which corresponds to a Carrington rotation period (Carrington 1859). A surface element on the Sun, whose image falls on pixel ij of the instrument detector, is at position (w_{ij}, n_{ij}, r_{ij}) relative to the centre of the Sun, where w is westward, n is northward and r is in the radial direction (see Thompson 2006 for more details on the coordinate system used). The spacecraft is at position $(0, 0, r_{sc})$. The w, n, r components of the spacecraft's position relative to each element ij can thus be written as:

---

[2]HMI data products can be downloaded online via the Joint Science Operations Center website: http://jsoc.stanford.edu.

5.3 Pixel Statistics from SDO/HMI Images

$$\delta w_{ij} = w_{ij} - 0$$
$$\delta n_{ij} = n_{ij} - 0 \quad (5.8)$$
$$\delta r_{ij} = r_{ij} - r_{sc}$$

The spacecraft's motion and the rotation of the Sun introduce velocity perturbations, which I determine in Sects. 5.3.1 and 5.3.2, respectively. These two contributions are then subtracted from each doppler image, thus revealing the Sun's magnetic activity velocity signatures. I compute the RV variations due to the suppression of convective blueshift and the flux blocked by sunspots on the rotating Sun in Sects. 5.3.5.3 and 5.3.5.4. I show that the Sun's activity-driven RV variations are well reproduced by a scaled sum of these two contributions in Sect. 5.4. Finally, I compute the disc-averaged magnetic flux and compare it as an RV proxy against the traditional spectroscopic activity indicators in Sect. 5.5.

5.3.1 Spacecraft Motion

The w, n, r components of the velocity incurred by the motion of the spacecraft relative to the Sun, $\mathbf{v}_{sc}$, are given in the fits header of each SDO/HMI observation. I normalise $\mathbf{v}_{sc}$ to account for variations in the spacecraft's position relative to the Sun. The magnitude of the spacecraft's velocity away from pixel ij can therefore be expressed as:

$$v_{sc,ij} = -\frac{\delta w_{ij}\, v_{sc,w_{ij}} + \delta n_{ij}\, v_{sc,n_{ij}} + \delta r_{ij}\, v_{sc,r_{ij}}}{d_{ij}}, \quad (5.9)$$

where:

$$d_{ij} = \sqrt{\delta w_{ij}^2 + \delta n_{ij}^2 + \delta r_{ij}^2} \quad (5.10)$$

is the distance between pixel ij and the spacecraft. I note that all relative velocities in this chapter follow the natural sign convention that velocity is rate of change of distance.

5.3.2 Solar Rotation

The solar rotation as a function of latitude was measured by Snodgrass and Ulrich (1990) in low resolution full-disc dopplergrams and magnetograms obtained at the Mount Wilson 150 foot tower telescope between 1967 and 1987. By cross-correlating time series of dopplergrams and magnetograms, they were able to determine the rate of motion of surface features (primarily supergranulation cells and sunspots) and deduce the rate of rotation of the Sun's surface as a function of latitude. The solar

Table 5.1 Solar differential rotation profile parameters from Snodgrass and Ulrich (1990)

Parameter	Value (deg day^{-1})
A	14.713
B	−2.396
C	−1.787

differential rotation profile $w(\phi)$ at each latitude ϕ is commonly described by a least squares polynomial of the form:

$$w(\phi) = A + B \sin^2 \phi + C \sin^4 \phi. \tag{5.11}$$

The best fit parameters found by Snodgrass and Ulrich (1990), used in this analysis, are given in Table 5.1. I apply this rotation profile in the heliographic frame to determine the w, n, r components induced by the solar rotation velocity along the line of sight to a given image pixel, $v_{\text{rot},w}$, $v_{\text{rot},n}$ and $v_{\text{rot},r}$. Normalising again by d, I can write:

$$v_{\text{rot}} = -\frac{\delta w \, v_{\text{rot},w} + \delta n \, v_{\text{rot},n} + \delta r \, v_{\text{rot},r}}{d}. \tag{5.12}$$

5.3.3 Flattened Continuum Intensity

I flatten the continuum intensity images using a fifth order polynomial function L_{ij} with the limb darkening constants given in Astrophysical Quantities (Allen 1973), through the IDL subroutine *darklimb_correct.pro*.[3] The flattened and non-flattened continuum intensities are related via the limb-darkening function L as follows:

$$I_{\text{flat},ij} = \frac{I_{ij}}{L_{ij}}. \tag{5.13}$$

5.3.4 Unsigned Longitudinal Magnetic Field Strength

The SDO/HMI instrument measures the line-of-sight (longitudinal) magnetic field strength B_{obs}. The magnetic field of the Sun stands radially out of the photosphere with a strength B_r. Due to foreshortening, the observed (longitudinal) field B_{obs} is less than the true (radial) field by a factor:

$$\mu_{ij} = \cos \theta_{ij}, \tag{5.14}$$

---

[3] Source code available at: http://hesperia.gsfc.nasa.gov/ssw/gen/idl/solar/.

5.3 Pixel Statistics from SDO/HMI Images

where θ_{ij} is the angle between the outward normal to the feature on the solar surface and the direction of the line-of-sight of the SDO spacecraft.

I can thus recover the full magnetic field strength by dividing by μ_{ij}:

$$B_{r,ij} = B_{\mathrm{obs},ij}/\mu_{ij}. \tag{5.15}$$

As is routinely done in solar work, I do not apply this operation for pixels that are very close to the limb ($\mu_{ij} < 0.1$) as it would lead me to overestimate the magnetic field strength.

The noise level in HMI magnetograms is a function of μ (Yeo et al. 2013). It is lowest for pixels in the centre of the CCD, where it is close to 5 G, and increases towards the edges and reaches 8 G at the solar limb. For this analysis I assume that the noise level is constant throughout the image with a conservative value $\sigma_{B_{\mathrm{obs},ij}} = 8$ G, in agreement with the results of Yeo et al. (2013). I therefore set $B_{\mathrm{obs},ij}$ and $B_{r,ij}$ to 0 for all pixels with a longitudinal field measurement ($B_{\mathrm{obs},ij}$) below this value.

5.3.5 Surface Markers of Magnetic Activity

5.3.5.1 Identifying Quiet-Sun Regions, Faculae and Sunspots

The first three panels of Fig. 5.3 show an SDO/HMI flattened intensitygram, line-of-sight Dopplergram and unsigned radial magnetogram for a set of images taken on 2011, November 10, after removing the contributions from spacecraft motion and solar rotation. I identify quiet-Sun regions, faculae and sunspots by applying magnetic and intensity thresholds.

- *Magnetic threshold:* The distribution of pixel unsigned observed magnetic field strength as a function of pixel flattened intensity is shown in Fig. 5.4. In the top histogram and main panel, we see that the distribution of magnetic field strength falls off sharply with increasing field strength. The vast majority of pixels are clustered close to 0 G: these pixels are part of the quiet-Sun surface. I separate active regions from quiet-Sun regions by applying a threshold in unsigned radial magnetic field strength for each pixel. Yeo et al. (2013) investigated the intensity contrast between the active and quiet photosphere using SDO/HMI data, and found an appropriate cutoff at:

$$|B_{r,ij}| > 3\,\sigma_{B_{\mathrm{obs},ij}}/\mu_{ij}, \tag{5.16}$$

where $\sigma_{B_{\mathrm{obs},ij}}$ represents the magnetic noise level in each pixel (see last paragraph of Sect. 5.3.4). As in Yeo et al. (2013), I exclude *isolated* pixels that are above this threshold as they are likely to be false positives. I can thus write:

$$|B_{r,\mathrm{thresh},ij}| = 24\,\mathrm{G}/\mu_{ij}. \tag{5.17}$$

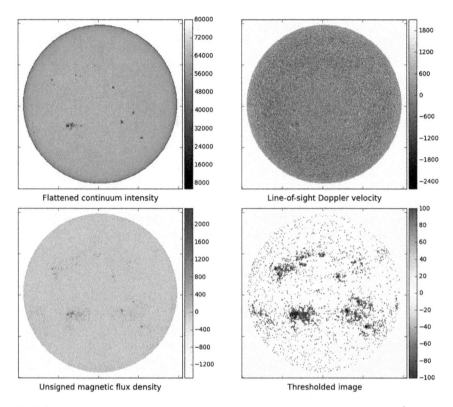

Fig. 5.3 *First three panels* SDO/HMI flattened intensity, line-of sight velocity (km s^{-1}) for the non-rotating *Sun* and unsigned longitudinal magnetic flux $|B_{\mathrm{l}}|/\mu$ (G) of the Sun, observed on 2011, November 10 at 00:01:30 UTC. *Last panel* my thresholded image, highlighting faculae (*blue pixels*) and sunspots (*red pixels*)

- *Intensity threshold:* The distribution of line-of-sight velocity as a function of pixel flattened intensity is shown in Fig. 5.5. The main panel allows us to further categorise active-region pixels into faculae and sunspots (umbra and penumbra). I apply the intensity threshold of Yeo et al. (2013):

$$I_{\mathrm{thresh}} = 0.89\, \hat{I}_{\mathrm{quiet}}, \tag{5.18}$$

where $\hat{I}_{\mathrm{quiet}}$ is the mean pixel flattened intensity over quiet-Sun regions. It can be calculated by summing the flattened intensity of each pixel that has $|B_{\mathrm{r},ij}| < |B_{\mathrm{r,thresh},ij}|$:

$$\hat{I}_{\mathrm{quiet}} = \frac{\sum_{ij} I_{\mathrm{flat},ij}\, W_{ij}}{\sum_{ij} W_{ij}}, \tag{5.19}$$

5.3 Pixel Statistics from SDO/HMI Images

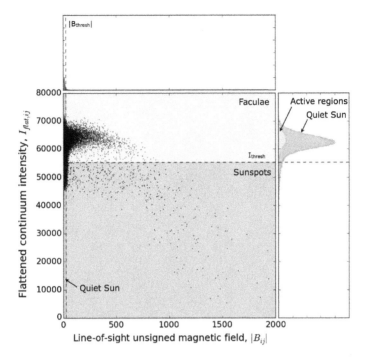

Fig. 5.4 Pixel line-of-sight (*longitudinal*) magnetic field strength, $|B_{\mathrm{obs},ij}|$, as a function of flattened intensity $I_{\mathrm{flat},ij}$, for the Sun on 2011, November 10 at 00:01:30 UTC. The *top* and *right* histograms show the distributions of $|B_{\mathrm{obs},ij}|$ and $I_{\mathrm{flat},ij}$, respectively. The *dashed lines* represent the cutoff criteria selected to define the quiet photosphere, faculae and sunspots. Over 95 % of the solar disc is magnetically quiet

where the weighting factors are defined as:

$$W_{ij} = 1 \text{ if } |B_{\mathrm{r},ij}| > |B_{\mathrm{r,thresh},ij}|,$$
$$W_{ij} = 0 \text{ if } |B_{\mathrm{r},ij}| < |B_{\mathrm{r,thresh},ij}|. \tag{5.20}$$

In the main panel of Fig. 5.5, quiet-Sun pixels are plotted in black, while active-region pixels are overplotted in yellow.

The last panel of Fig. 5.3, which shows the thresholded image according to these $I_{\mathrm{flat},ij}$ and $|B_{\mathrm{r},ij}|$ criteria, confirms that they are effective at identifying sunspot and faculae pixels correctly.

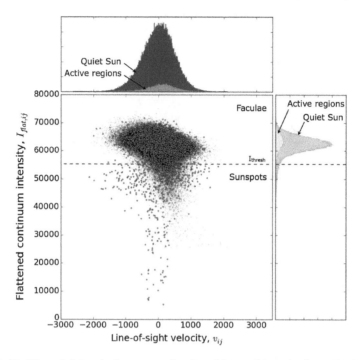

Fig. 5.5 Pixel line-of-sight velocity, v_{ij}, as a function of flattened intensity $I_{\text{flat},ij}$, for the Sun on 2011, November 10 at 00:01:30 UTC. The *top* and *right* histograms show the distributions of v_{ij} and $I_{\text{flat},ij}$, respectively. In the case of active pixels (*yellow dots*), the line of sight velocity is invariant with pixel brightness. For quiet-Sun pixels (*black dots*), however, brighter pixels are *blueshifted* while fainter pixels are *redshifted*: this effect arises from granular motions

5.3.5.2 Velocity Contribution of Convective Motions in Quiet Sun Regions

I estimate the average RV of the quiet Sun by summing the intensity-weighted velocity of non-magnetised pixels, after removing the spacecraft motion and the Sun's rotation:

$$\hat{v}_{\text{quiet}} = \frac{\sum_{ij}(v_{ij} - \delta v_{\text{sc},ij} - \delta v_{\text{rot},ij})\, I_{ij}\, W_{ij}}{\sum_{ij} I_{ij}\, W_{ij}}. \tag{5.21}$$

For this calculation, I define the weights as follows:

$$\begin{aligned} W_{ij} &= 1 \text{ if } |B_{\text{r},ij}| < |B_{\text{r,thresh},ij}|, \\ W_{ij} &= 0 \text{ if } |B_{\text{r},ij}| > |B_{\text{r,thresh},ij}|. \end{aligned} \tag{5.22}$$

This velocity field is thus averaged over the vertical motions of convection granules on the solar surface. Hot and bright granules rise up to the surface, while cooler and darker fluid sinks back towards the Sun's interior. This process is visible in the main panel of Fig. 5.5: quiet-Sun pixels (black dots) are clustered in a tilted ellipse. The area of the upflowing granules is larger than that enclosed in the intergranular lanes, and the granules are carrying hotter and thus brighter fluid. This results in a net blueshift, as seen in Fig. 5.5.

5.3.5.3 Suppression of Convective Blueshift from Active Regions

The presence of magnetically active regions inhibits convection and therefore acts to suppress this blueshift. I measure the total disc-averaged velocity of the Sun $\hat{v}$ by summing the velocity contribution of each pixel ij, weighted by their intensity I_{ij}, after subtracting the spacecraft motion and solar rotation:

$$\hat{v} = \frac{\sum_{ij}(v_{ij} - \delta v_{\text{sc},ij} - \delta v_{\text{rot},ij})\, I_{ij}}{\sum_{ij} I_{ij}} \quad (5.23)$$

The suppression of granular blueshift induced by magnetically active regions ($|B_{\text{r},ij}| > |B_{\text{r,thresh},ij}|$) is therefore:

$$\Delta \hat{v}_{\text{conv}} = \hat{v} - \hat{v}_{\text{quiet}}. \quad (5.24)$$

The value of $\Delta \hat{v}_{\text{conv}}$ at each time of the HARPS observations is listed in the Appendix table of Haywood et al. (2016).

5.3.5.4 Rotational Perturbation Due to Sunspot Flux Deficit

As the Sun rotates, the presence of dark spots on the solar surface breaks the Doppler balance between the approaching (blueshifted) and receding (redshifted) hemispheres. The resultant velocity perturbation can be obtained by summing the line-of-sight velocity of sunspot pixels corrected for the spacecraft's motion, and weighted by the deficit in flux produced by the presence of a sunspot:

$$\Delta \hat{v}_{\text{spots}} = \frac{\sum_{ij}(v_{ij} - \delta v_{\text{sc},ij})(I_{ij} - L_{ij})\, W_{ij}}{\sum_{ij} I_{ij}} \quad (5.25)$$

In this case, the weights are set to 1 only for pixels that fulfill both the magnetic strength and brightness criteria:

$$W_{ij} = 1 \text{ if } |B_{\text{r},ij}| > |B_{\text{r,thresh},ij}|$$
$$\text{and} \quad (5.26)$$
$$I_{\text{flat},ij} < 0.89\, \hat{I}_{\text{quiet}}.$$

5.4 Reproducing the RV Variations of the Sun

In this section, I combine our model of Vesta's rotational RV signal (presented in Sect. 5.2.4.1) with the two magnetic activity basis functions determined in Sects. 5.3.5.3 and 5.3.5.4, in order to reproduce the RV variations seen in the HARPS observations.

5.4.1 Total RV Model

The final model has the form:

$$\Delta RV_{\text{model}}(t) = A \, \Delta \hat{v}_{\text{conv}}(t) + B \, \Delta \hat{v}_{\text{spots}}(t) + \Delta RV_{\text{vesta}}(t) + RV_0. \quad (5.27)$$

I carry out an optimal scaling procedure in order to determine the scaling factors (A, B, C and S) of each of the contributions, as well as the constant offset RV_0. Each basis function is orthogonalised by subtracting its inverse-variance weighted average prior to performing the scaling. I determine the maximum likelihood via a procedure similar to the one described in Collier Cameron et al. (2006). This procedure is applied to the unbinned (not nightly-averaged) HARPS dataset, in order to determine the appropriate scaling coefficients (C and S) for Vesta's axial rotation. The total amplitude of the modulation induced by Vesta's rotation is equal to 2.39 m s^{-1}, which is of the same order as the amplitude I estimated in Sect. 5.2.4.1. After all the scaling coefficients were determined, I grouped the observations in each night by computing the inverse variance-weighted average for each night. The final model is shown in Fig. 5.6, and the best-fit values of the parameters are listed in Table 5.2.

Panel (e) shows the residuals remaining after subtracting the total model ΔRV_{model} from the HARPS observations of the Sun as-a-star ΔRV_{Sun}. The first part of the run (nights 0–11) displays a residual rms of 3.72 m s^{-1}, while the second part (nights 36–68) has an rms of 1.38 m s^{-1}. As I mentioned in Sect. 5.2.4.2, I attribute the excess scatter in the first nights to 2–3 h granulation signals that were not well-averaged with our observing strategy. The observing strategy deployed in the second part of the run appears to be much more effective at mitigating granulation signals, even though a few outliers do remain (e.g., at night 52). They may be affected by super-granulation motions which commonly take place over timescales of 8 h or longer, and which could result in residual white noise from one night to the next.

5.4 Reproducing the RV Variations of the Sun

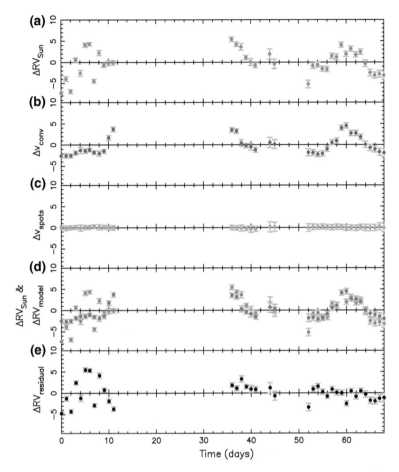

Fig. 5.6 *Panel* **a** HARPS RV variations of the Sun as-a-star, ΔRV_{Sun}; *Panel* **b** Scaled basis function for the suppression of convective *blueshift*, $\Delta \hat{v}_{conv}$, derived from SDO/HMI images; *Panel* **c** Scaled basis function for the rotational perturbation due to sunspot flux deficit, $\Delta \hat{v}_{rot}$; *Panel* **d** total RV model, ΔRV_{model} (*red*), overlaid on top of the HARPS RV variations (*blue points*); *Panel* **e** residuals obtained after subtracting the model from the observations. All RVs are in m s^{-1}. Note that the scale of the y-axis is different to that used in Fig. 5.2

5.4.2 Relative Importance of Suppression of Convective Blueshift and Sunspot Flux Deficit

We see that the activity-induced RV variations of the Sun are well reproduced by a scaled sum of the two basis functions, $\hat{v}_{conv}$ and $\hat{v}_{spots}$ (shown in panels (b) and (c), respectively). As previously predicted by (Meunier et al. 2010), I find that the suppression of convective blueshift plays a dominant role (RMS of 2.22 m s^{-1}). I also found this to be the case for CoRoT-7, a main sequence G9 star with a rotation period

Table 5.2 Best-fit parameters resulting from the optimal scaling procedure

Parameter	Value
A	0.64 ± 0.29
B	2.09 ± 0.06
C	1.99 ± 0.08
S	1.33 ± 0.09
RV_0 (m s^{-1})	99.80 ± 2.90

comparable to that of the Sun (see Chap. 4, Sect. 4.1 and Haywood et al. (2014)). The relatively low amplitude of the modulation induced by sunspot flux-blocking (rms of 0.14 m s^{-1}) is expected in slowly-rotating stars with a low $v \sin i$ (Desort et al. 2007). As the suppression of convective blueshift by active regions clearly dominates the total activity-induced RV variations of the Sun, I did not compute the RV modulation induced by facular flux-brightening; this contribution would only be a second-order effect.

5.4.3 Zero Point of HARPS

The wavelength adjustments that were applied to the HARPS RVs were based on precise prior dynamical knowledge of the rate of change of distance between the Earth and Vesta, and between Vesta and the Sun. The offset $RV_0 = 99.80 \pm 2.90$ m s^{-1} thus represents the zero point of the HARPS instrument, including the mean granulation blueshift for the Sun.

5.5 Towards Better Proxies for RV Observations

5.5.1 Disc-Averaged Observed Magnetic Flux $|\hat{B}_{\mathrm{obs}}|$

The averaged magnetic flux may be a useful proxy for activity-driven RV variations as it should map onto areas of strong magnetic fields, which suppress the Sun's convective blueshift. The line-of-sight magnetic flux density and filling factor on the visible hemisphere of a star can be measured from the Zeeman broadening of magnetically-sensitive lines (Robinson 1980; Reiners et al. 2013). Their product gives the disc-averaged flux density that we are deriving from the solar images.

I compute the full-disc line-of-sight magnetic flux of the Sun, by summing the intensity-weighted line-of-sight unsigned magnetic flux in each pixel:

$$|\hat{B}_{\mathrm{obs}}| = \frac{\sum_{ij} |B_{\mathrm{obs},ij}| I_{ij}}{\sum_{ij} I_{ij}} \qquad (5.28)$$

5.5 Towards Better Proxies for RV Observations

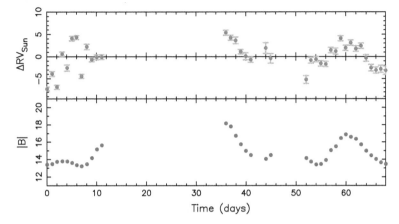

Fig. 5.7 *Top* HARPS RV variations of the Sun as-a-star (m s^{-1}); *bottom* variations of the disc-averaged line-of-sight magnetic flux $|\hat{B}_{\mathrm{obs}}|$ (G). The two follow each other closely

The variations in $|\hat{B}_{\mathrm{obs}}|$ are shown in Fig. 5.7, together with the nightly-averaged HARPS RV variations of the Sun as-a-star. We see that the variations in the disc-averaged magnetic flux are in phase with the RV variations, despite the scatter in RV in the first part of the run (discussed in Sect. 5.2.4.2).

5.5.2 Correlations Between RV and Activity Indicators

Figure 5.8 presents the correlations between the nightly-averaged HARPS RV variations of the Sun as-a-star, the activity basis functions $\hat{v}_{\mathrm{conv}}$ and $\hat{v}_{\mathrm{spots}}$ and the full-disc magnetic flux computed from the SDO/HMI images, the observed FWHM, BIS, and $\log(R'_{\mathrm{HK}})$ derived from the HARPS DRS reduction pipeline. I computed the Spearman correlation coefficient to get a measure of the degree of monotone correlation between each variable (the correlation between two variables is not necessarily linear, for example between RV and BIS). The coefficients are displayed in each panel of Fig. 5.8, both including and excluding the observations made in the first part of the run, which show a lot of intra-night scatter. Although the extra scatter seen in the first block of observations does affect the trend slightly, it is clear that the activity-induced RV variations of the Sun are significantly correlated with the disc-averaged magnetic flux. If I only consider the observations in the second part of the run, the Spearman correlation coefficient between the RV variations of the Sun as-a-star and the disc-averaged magnetic flux is equal to 0.83. The correlation is stronger between $|\hat{B}_{\mathrm{obs}}|$ and $\hat{v}_{\mathrm{conv}}$, with a correlation coefficient of 0.89, which is in agreement with the fact that magnetised areas suppress convective blueshift. The RV variations due to sunspot flux deficit are not significantly correlated with the disc-averaged magnetic flux (or with any of the other activity indicators), but this is not so critical since these

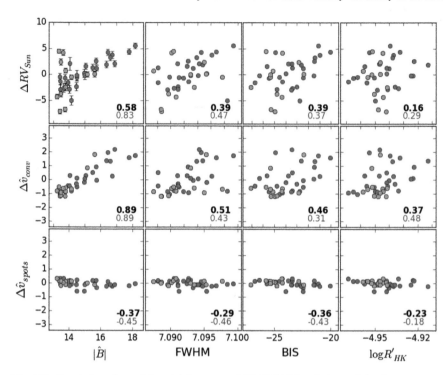

Fig. 5.8 Correlation plots of the nightly-averaged HARPS RV variations of the Sun as-a-star, suppression of convective blueshift $\Delta \hat{v}_{\mathrm{conv}}$, and modulation due to sunspot flux deficit $\Delta \hat{v}_{\mathrm{spots}}$ against (from *left* to *right*): the disc-averaged observed magnetic flux $|\hat{B}_{\mathrm{obs}}|$ (G), FWHM (km s^{-1}), BIS (m s^{-1}) and $\log(R'_{\mathrm{HK}})$. Observations from the first part of the *run* are highlighted in a *lighter* shade. Spearman correlation coefficients are displayed in the *bottom-right corner* of each *panel*: for the full observing run (in *bold* and *black*), and for the second part of the run only (in *blue*)

variations only play a minor role in the total activity-induced RV variations of the Sun. When compared against correlations with the traditional spectroscopic activity indicators (the FWHM, BIS and $\log(R'_{\mathrm{HK}})$), I see that the disc-averaged magnetic flux $|\hat{B}_{\mathrm{obs}}|$ is a much more effective proxy for activity-induced RV variations.

5.6 Summary

In this chapter, I decomposed activity-induced RV variations into identifiable contributions from sunspots, faculae and granulation, based on Sun as-a-star RV variations deduced from HARPS spectra of the bright asteroid Vesta and high spatial resolution images taken with the Helioseismic and Magnetic Imager (HMI) instrument aboard the Solar Dynamics Observatory (SDO). I find that the RV variations induced by solar activity are mainly caused by the suppression of convective blueshift from magnetically active regions, while the flux deficit incurred by the presence of sunspots on the rotating solar disc only plays a minor role. I further compute the disc-averaged line-of-sight magnetic flux and show that it is an excellent proxy for activity-driven RV variations, much more so than the full width at half-maximum and bisector span of the cross-correlation profile, and the Ca II H&K activity index.

In addition to the existing 2011 HARPS observations of sunlight scattered off Vesta, there will soon be a wealth of direct solar RV measurements taken with HARPS-N, which will be regularly fed sunlight through a small 2-inch telescope developed specifically for this purpose. A prototype for this is currently being commissioned at HARPS-N (Glenday et al., in prep.). Gaining a deeper understanding of the physics at the heart of activity-driven RV variability will ultimately enable us to better model and remove this contribution from RV observations, thus revealing the planetary signals.

In the future, I wish to take this investigation one step further by synthesizing Sun-as-a-star CCFs, using SDO/HMI continuum and Dopplergram images, which contain information on the intensity scale and velocity shift of each pixel of the Sun. This will reveal spectral line profile distortions produced by activity. Comparing these synthetic line profiles with the observed HARPS CCFs (plotted in Fig. 2.5 of Chap. 2) will provide a unique insight on the physical processes at play in magnetic RV variability.

References

Allen C (1973) Allen: astrophysical quantities, 3rd edn. The Athlone Press, University of London
Baranne A et al (1996) Astron Astrophys Suppl Ser 119:373
Boisse I, Bouchy F, Hébrard G, Bonfils X, Santos N, Vauclair S (2011) Astron Astrophys 528:A4
Boisse I et al (2009) Astron Astrophys 495:959
Bonfils X et al (2007) Astron Astrophys 474:293
Carrington RC (1859) Month Not R Astron Soc 19:81
Collier Cameron A et al (2006) Month Not R Astron Soc 373:799
Deming D, Espenak F, Jennings DE, Brault JW, Wagner J (1987) Astrophys J 316:771
Deming D, Plymate C (1994) Astrophys J 426:382
Desort M, Lagrange AM, Galland F, Udry S, Mayor M (2007) Astron Astrophys 473:983
Gregory PC (2011) MNRAS 415:2523 1101.0800
Haywood RD et al (2014) Monthly notices of the royal astronomical society. 443(3):2517–2531
Haywood RD et al (2016) arXiv:1601.05651
Huélamo N et al (2008) Astron Astrophys 489:L9
Jiménez A, Pallé PL, Régulo C, Roca Cortes T, Isaak GR (1986) COSPAR and IAU. 6–89
Lagrange AM, Desort M, Meunier N (2010) Astron Astrophys 512:A38
Lanza AF, Molaro P (2015) Experimental astronomy. 39:461. 1505.00918
Lindegren L, Dravins D (2003) Astron Astrophys 401:1185
Lovis C, Pepe F (2007) Astron Astrophys 468:1115
Makarov VV, Beichman CA, Catanzarite JH, Fischer DA, Lebreton J, Malbet F, Shao M (2009) Astrophys J 707:L73
Meunier N, Desort M, Lagrange AM (2010) Astron Astrophys 512:A39
Molaro P, Centurión M (2010) Astron Astrophys 525:A74

McMillan RS, Moore TL, Perry ML, Smith PH (1993) ApJ 403:801–809, doi:10.1086/172251, http://adsabs.harvard.edu/abs/1993ApJ...403..801M
Pesnell WD, Thompson BJ, Chamberlin PC (2012) Sol Phys 275:3
Queloz D et al (2001) Astron Astrophys 379:279
Reiners A, Shulyak D, Anglada-Escudé G, Jeffers SV, Morin J, Zechmeister M, Kochukhov O, Piskunov N (2013) Astron Astrophys 552:103
Robertson P, Mahadevan S, Endl M, Roy A (2014) Science 345:440
Robinson RD Jr (1980) APJ 239:961
Russell CT et al (2012) Science 336:684
Santos NC et al (2014) Astron Astrophys 566:35
Schou J et al (2012) Sol Phys 275:229
Snodgrass HB, Ulrich RK (1990) Astrophys J 351:309
Stephenson CB (1951) Astrophys J 114:500
Thomas PC, Binzel RP, Gaffey MJ, Zellner BH, Storrs AD, Wells E (1997) Icarus 128:88
Thompson WT (2006) Astron Astrophys 449:791
Yeo KL, Solanki SK, Krivova NA (2013) A&A. 550:A95 1302.1442

Chapter 6
Conclusion: Next Steps and Aims for the Future

Thousands of exoplanets have now been found, the majority of which were discovered or confirmed after follow-up with RV observations. Spectrographs such as the 3.6 m/HARPS and TNG/HARPS-N are capable of measuring RVs of bright stars with sub-metre per second precision. The intrinsic variability of the stars themselves, however, currently remains the main obstacle to determining the masses of small planets. The presence of magnetic features on the stellar surface, such as starspots, faculae/plage and granulation, can induce quasi-periodic RV variations of over several metres per second, which can easily conceal the orbits of super-Earths and Earth-mass planets.

I developed a Monte Carlo Markov Chain code that detects exoplanet orbits in the presence of stellar activity, which I presented in Chap. 3. Activity-induced RV signals are intimately tied to the star's rotation period, and their frequency structure is governed by the constantly-evolving magnetic features on the stellar surface. I modelled the correlated noise arising from the star's magnetic activity using a Gaussian process that has the same covariance function, or frequency structure, as the off-transit variations in the star's lightcurve. This new activity decorrelation technique allows me to identify the orbital signatures of planets present in a system and to determine their masses, with realistic allowance for the uncertainty introduced by the stellar activity. I implemented state of the art Bayesian model comparison tools to avoid over-fitting and determine the number of planets present in a system.

I applied my code to several high-precision RV datasets, as reported in Chap. 4. I analysed the simultaneous 3.6 m/HARPS RVs and CoRoT photometric time series of the active star CoRoT-7, host to a transiting super-Earth and a small Neptune, which has been the subject of much debate in recent years due to its high activity levels. I also determined the masses of Kepler-10b and c using HARPS-N RV observations, and of Kepler-78b by combining the HARPS-N and HIRES RV datasets together.

In parallel, I studied the Sun in order to gain a deeper understanding of the processes at the heart of activity-driven RV signals, as described in Chap. 5. The Sun is the only star for which we can resolve individual surface structures that are the source of stellar RV variability. I used high spatial resolution SDO/HMI continuum, Dopplergram and magnetogram images to determine the RV signatures of sunspots, faculae/plage and granulation. I also determined the Sun's total RV variations over two solar rotations using 3.6 m/HARPS observations of sunlight scattered by the surface of the bright asteroid Vesta. I tested these variations against the RV contribution determined from the SDO/HMI images and found that the activity-driven RV variations of the Sun are strongly correlated with its full-disc magnetic flux. This result may become key to disentangling planetary orbits from stellar activity in future years.

Next Steps

The detection and characterisation of exoplanets is a very dynamic and fast-moving field. The stellar activity barrier is one of the main challenges faced by the exoplanet community today, and we must overcome this barrier in order to become able to routinely detect Earth-mass planets at larger distances from their stars.

I now intend to tackle this issue via a two-fold approach:

- Incorporate activity proxies into my existing framework based on Gaussian processes and Bayesian model selection;
- Explore the temporal behaviour and physical origin of the magnetic processes at the heart of stellar RV variability, through the study of *Kepler* stars and the Sun.

An intuitive and rigorous approach to modelling RV stellar variability Long term, high precision photometry such as was obtained during the *Kepler* and CoRoT missions, is not available for the majority of candidates selected for RV follow-up. TESS will only provide us with around 30 days of photometry, which will be too short to capture fully the activity patterns of stars on their rotation and magnetic activity timescales. I wish to use Gaussian processes to develop robust activity RV models based on spectroscopic indicators (the bisector and full width at half maximum of the cross-correlation function, the R'_{HK} index) as well as new diagnostics derived from large-scale MHD simulations of photospheric convection (eg., Cegla et al. 2013). Their frequency structure is similar to that of the intrinsic magnetic activity of the host star, and can be encoded within the covariance function of a Gaussian process. Furthermore, stellar activity signals are quasi-periodic in nature, whereas planet orbits are fully periodic. The Gaussian process framework provides a means to identify a truly coherent and periodic signal, when implemented in parallel with a robust model comparison tool. I wish to test my models in a systematic way using synthetic datasets to assess the detectability of planets in the presence of stellar activity. This will help to identify the most promising targets for RV follow-up of *Kepler* and K2 candidates, and to devise observing strategies that will further minimise the impact of stellar activity, in readiness for the TESS, CHEOPS and JWST missions.

Deciphering magnetic activity patterns on the stellar rotation timescale As well as exploring individual systems, I wish to undertake a large-scale study of the activity patterns of Sun-like stars to look for relations between their photometric and RV variability, over stellar rotation timescales (as a continuation of the work I presented in Chap. 2, Sect. 2.3). The Fourier components of the lightcurve provide important clues about the complexity of the activity-induced RV variations (Bastien et al. 2014). In this perspective, decoding the temporal structure of a star's lightcurve is a natural step towards understanding stellar RV variability. I wish to find out whether certain groups of stars (eg., for a given spectral type, or age) display a distinct kind of magnetic activity behaviour. It is already known that young stars tend to show spot-dominated photometric variability, whereas old stars are faculae-dominated (Radick et al. 1983, 1987, 1995; Lockwood et al. 2007). I wish to explore the dependancy of spectral type on the links between the photometric RMS, rotation period and shape of the autocorrelation function of the lightcurve in main-sequence stars (spanning the late F to early M spectral classes). These parameters can be easily obtained from *Kepler* light curves by applying autocorrelation and Lomb-Scargle periodogram techniques, which I have already implemented (see Chap. 2, Sects. 2.3.2.1 and 2.3.2.2). Studying the lifetimes and sizes of starspot regions may also allow me to identify different types of magnetic activity behaviour. Classifying stars depending on their activity behaviour will allow the exoplanet community to develop better tailored models to account for RV variability, and may also help to pick more "manageable" stars in future RV surveys. This work will also enhance our understanding of stellar surface details, magnetic fields, and how they vary with mass and age/rotation.

Probing the physics at the heart of the sun's RV variability I plan to pursue my current study of the Sun to develop the next generation of more sophisticated activity models. In addition to the existing 2012 HARPS observations of sunlight reflected off Vesta, there will soon be a wealth of direct solar RV measurements taken with HARPS-N, which will be regularly fed sunlight through a small 2-in. telescope developed specifically for this purpose (Dumusque et al. 2015; Glenday et al. in prep.). A prototype for this is currently being commissioned at HARPS-N. In particular, I wish to explore the effect of faculae on the suppression of convective blueshift, since this process has been found to be the dominant contribution to the activity-induced RV signal (Meunier et al. 2010a, b; Haywood et al. 2014). Other types of photospheric velocity field may play an important but previously unrecognised role in stellar RV variability; in particular, Gizon et al. (2010, 2001) report the presence of ~ 50 m $\cdot$ s^{-1} inflows towards active regions on the Sun's surface. Planetary signals are the same at all wavelengths, whereas stellar activity signals will change according to the photospheric depth sampled by different line masks of different wavelength ranges Anglada-Escudé and Butler (2012); Tuomi and Anglada-Escudé (2013). I wish to explore the physical sources of this phenomenon, and investigate the possibility of incorporating the information gained from this wavelength dependance into my code. Gaining a deeper understanding of the physics at the heart of activity-driven RV variability will ultimately enable us to better model and remove this contribution from RV observations, thus revealing the planetary signals.

References

Anglada-Escudé G, Butler RP (2012) ApJS 200:15. arXiv:1202.2570
Bastien FA et al (2014) Astron J 147:29
Cegla HM, Shelyag S, Watson CA, Mathioudakis M (2013) Astrophys J 763:95
Dumusque X et al (2015) ApJ Lett 814:L21. arXiv:511.02267
Gizon L, Duvall TLJ, Larsen RM (2001) Proc Int Astron Union 203:189
Gizon L, Birch AC, Spruit HC (2010) Ann Rev Astron Astrophys 48:289
Glenday A, Phillips DF et al. in prep
Haywood RD et al (2014) Monthly Not R Astron Soc 443(3):2517–2531, 443, 2517
Lockwood GW, Skiff BA, Henry GW, Henry S, Radick RR, Baliunas SL, Donahue RA, Soon W (2007) Astrophys J Suppl Ser 171:260
Meunier N, Desort M, Lagrange AM (2010a) Astron. Astrophys 512:A39
Meunier N, Lagrange A-M, Desort M (2010b) A&A 519:A66. arXiv:1005.4764
Radick RR, Lockwood GW, Skiff BA, Thompson DT (1995) Astrophys J 452:332
Radick RR, Thompson DT, Lockwood GW, Duncan DK, Baggett WE (1987) Astrophys J 321:459
Radick RR et al (1983) PASP 95:300
Tuomi M, Anglada-Escudé G (2013) Astron Astrophys 556:A111

Index

A
Autocorrelation, 34, 38, 54, 55, 94, 101
Autocorrelation functions, 36–39, 54, 55, 79, 94, 102, 103, 137

B
Blueshift, 25

C
Chromosphere, 14, 21, 22
Convective blueshift, 22, 31, 32, 41, 60, 78, 79, 83, 87, 114, 121, 127, 129–132, 137
Covariance functions, 53–58, 60, 78–83, 88, 91–94, 102, 103, 105, 135, 136
Covariance matrix, 50–53, 56–58, 93
Cross-correlation functions (CCFs), 8, 20, 27, 75, 76, 88, 133, 136
Cross-correlation profile, 40, 133

F
Faculae, 14, 18, 20–25, 31, 32, 41, 75, 88, 113, 114, 120, 123–125, 132, 135–137
Facular flux-brightening, 130
FF', 61
FF' method, 31, 32, 41, 60, 78–81, 88
Flocculi, 21

G
Gaussian priors, 79, 81, 93, 103

Gaussian processes, 8, 45, 46, 50, 54–56, 58–61, 78–83, 85, 88, 89, 91–99, 102–105, 107–110, 136
Granulation, 6, 14–17, 22, 25, 28, 29, 38, 40, 95, 113, 114, 119, 120, 128, 130, 132, 135, 136

H
Hyperparameters, 53–56, 58, 61, 62, 64, 79, 91, 93, 94, 103

I
Indicators, 22, 23, 26, 27, 41, 73, 74, 121, 131, 136

J
Jeffreys, 56, 62, 63, 94

L
Likelihood, 5, 50, 52, 64–68, 82, 83, 128
Limb-brightened facular emission, 60, 78, 88
Limb-darkening, 31, 122

M
Magnetic fields, 13, 14, 18, 19, 21, 22, 31, 114, 122, 123, 125, 130
Magnetic flux, 18, 20, 113–115, 121, 124, 130–133, 136
Marginal likelihood, 67, 68, 81, 82
Modified Jeffreys, 56, 63

O
Oscillations, 5, 6, 13, 15, 25, 95, 119

P
Periodograms, 15, 17, 31, 34, 35, 37, 38, 55, 65, 77, 78, 83, 85, 86, 88, 95, 97, 103–105, 107, 137
Photosphere, 14, 17–19, 21, 22, 27, 31, 55, 79, 113, 122
Plage, 18, 21–23, 25, 135, 136
Proxies, 8, 25, 33, 110, 113, 115, 121, 130, 132, 133, 136

Q
Quiet photosphere, 22, 123, 125
Quiet Sun, 123–127

S
Starspots, 18–22, 24, 25, 27, 30–32, 40, 54, 55, 60, 73, 75, 78, 83, 87, 113, 135
Sunspots, 17–21, 23, 24, 113, 114, 119–121, 123–125, 127, 129–133, 136
Supergranulation, 114, 121

U
Uniform prior, 56, 60, 63

CPSIA information can be obtained
at www.ICGtesting.com
Printed in the USA
LVHW02s2232160718
583946LV00003B/15/P